Ana Maria Bratu

CO2 Laser Photoacoustic Spectroscopy for Ultrasensitive Gas Detection

AF190958

Ana Maria Bratu

CO2 Laser Photoacoustic Spectroscopy for Ultrasensitive Gas Detection

Applications of laser photoacoustic spectroscopy in trace gas measurements

LAP LAMBERT Academic Publishing

Imprint

Any brand names and product names mentioned in this book are subject to trademark, brand or patent protection and are trademarks or registered trademarks of their respective holders. The use of brand names, product names, common names, trade names, product descriptions etc. even without a particular marking in this work is in no way to be construed to mean that such names may be regarded as unrestricted in respect of trademark and brand protection legislation and could thus be used by anyone.

Cover image: www.ingimage.com

Publisher:
LAP LAMBERT Academic Publishing
is a trademark of
Dodo Books Indian Ocean Ltd. and OmniScriptum S.R.L publishing group

120 High Road, East Finchley, London, N2 9ED, United Kingdom
Str. Armeneasca 28/1, office 1, Chisinau MD-2012, Republic of Moldova, Europe
Managing Directors: Ieva Konstantinova, Victoria Ursu
info@omniscriptum.com

Printed at: see last page
ISBN: 978-3-659-63546-5

CO₂ Laser Photoacoustic Spectroscopy for Ultrasensitive Gas Detection

Applications of laser photoacoustic spectroscopy in trace gas measurements

Edited by Dr. Ana Maria Bratu

National Institute for Laser, Plasma and Radiation Physic, Laser Department, 409 Atomistilor St., PO Box MG-36, 077125 Bucharest, Romania

Contents

Preface

Over the last four decades, technological developments in the field of lasers and high-sensitivity pressure detection systems (microphones and electronics) have contributed to the substantial progress of photoacoustic spectroscopy. In gases, the photoacoustic spectroscopy is based on the detection of an acoustical signal generated in a gas excited by a modulated laser beam at a wavelength corresponding to a absorption line of the gas species. The applications of resonant photoacoustic spectroscopy include concentration measurements and trace gas analysis, accurate determinations of thermophysical properties, detections of dynamic processes such as gas mixing or chemical reactions, relaxation processes (determinations of collisional lifetimes of specified quantum states and routes of energy exchange in polyatomic molecules), spectroscopic experiments, studies of aerosols, etc. Trace-gas sensing is a rapidly developing field, in demand for applications such as process and air-quality measurements, atmospheric monitoring, breath diagnostics, biology and agriculture, chemistry, and security and workplace surveillance.

The favorable properties of laser photoacoustic spectroscopy are essentially determined by the characteristics of the laser. The kind and number of detectable substances is related to the spectral overlapping of the laser emission with the absorption bands of the trace gas molecules. Thus, the accessible wavelength range, tunability, and spectral resolution of the laser are of prime importance. In conjunction with line-tunable IR lasers e.g., CO$_2$ lasers, the in situ monitoring of many substances occurring at ppbV (parts per billion volume) or even pptV (parts per trillion volume) concentrations, the photoacoustic detection provides not only high sensitivity but also the necessary selectivity for analyzing multicomponent mixtures.

This book is an overview of various applications of LPAS method. The aim is to provide extremely detailed and informative data of a functional photoacoustic instrument with various applications in different disciplines.

Greater clarity on the principles of laser photoacoustic spectroscopy is important for the understanding the detection of trace gas concentration at sub ppbV level. The experimental measures for reducing the influence of interfering gases in a single component measurement are discussed. Precise measurements of the absorption coefficient of ethylene at CO$_2$ laser wavelengths are a prerequisite for applications in highly regulated fields. All of the parameters that characterize photoacoustic cells, including the limiting sensitivity of the system, are measured and compared with the best results reported by other authors. Approaches to improving the current sensor performance are also tackled. The main components are lengthily described based on a general schematic. Special consideration is given to the home-built, frequency-stabilized, line-tunable CO$_2$-laser source and the resonant photoacoustic cell. Other aspects of a functional photoacoustic instrument, such as the gas handling system and data acquisition and processing are outlined.

This book provides examples of applications for LPAS methods. Measuring human biomarkers in exhaled breath is expected to revolutionize diagnosis and management of many diseases and may soon lead to rapid, improved, lower-cost diagnosis, which will in turn ensure expanded life spans and an improved quality of life.

New investigations based on LPAS technique have been performed to analyze hazardous gases from surgical smoke. Analysis of surgical smoke is important in evaluation of risk to at which the medical staff and the patients are exposed.

Chapter 1.
CO$_2$ laser photoacoustic spectroscopy system

1.1. Introduction

The photoacoustic (PA) spectroscopy is based on the effect discovered by Graham Bell in 1880 and consists of the energy conversion of modulated light radiation to sound energy (Bell, 1880). A part of the energy absorbed by the sample under study is transformed into thermal energy due to the nonradiative transitions. The temperature variations determine the formations of acoustic waves that can be detected directly by appropriate sensors (sensitive microphones). The potential of laser photoacoustic spectroscopy has been discussed in several review articles (Patel and Tam, 1981; West, 1983; Hess, 1983; Tam, 1986; Sigrist, 1986, 2003; Meyer and Sigrist, 1990; Harren and Reuss, 1997; Harren et al., 2000; Miklos et al., 2001; Schmid, 2006) and books (Pao, 1977; Rosencwaig, 1980; Zharov and Letokhov, 1986; Hess, 1989; Mandelis, 1992, 1994; Bicanic, 1992; Gusev and Karabutov, 1993; Sigrist, 1994; Mandelis and Hess, 1997).

PA spectroscopy is an indirect technique in that an effect of absorption is measured rather than absorption itself. Hence the name of photoacoustic: light absorption is detected through its accompanying acoustic effect. The advantage of photoacoustics is that the absorption of light is measured on a zero background; this is in contrast with direct absorption techniques, where a decrease of the source light intensity has to be observed. The spectral dependence of absorption makes it possible to determine the nature of the trace components. The PA method is primarily a calorimetric technique, which measures the precise number of absorbent molecules by simply measuring the amplitude of an acoustic signal. In laser photoacoustic spectroscopy (LPAS) the nonradiative relaxation which generates heat is of primary importance. In the infrared (IR) spectral region, nonradiative relaxation is much faster than radiative decay. Infrared spectroscopy is an efficient method to detect various gaseous species since many of them absorb in this spectral range (Dumitras et al., 2007).

PA spectroscopy relies on the PA effect for the detection of absorbing analytes. The sample gas is in a confined (resonant or nonresonant) chamber, where modulated (e.g., chopped) radiation enters via an IR-transparent window and is locally absorbed by IR-active molecular species. The temperature of the gas thereby increases, leading to a periodic expansion and contraction of the gas volume synchronous with the modulation frequency of the radiation. This generates a pressure wave that can be

3

acoustically detected by a suitable sensor, e.g., by a microphone. The advantages of the PA method are high sensitivity and small sample volume; besides, the acoustic measurement makes optical detection unnecessary. The main drawback is caused by the sensitivity to acoustic noise, because the measurements are based on an acoustic signal.

The favorable properties of LPAS are essentially determined by the characteristics of the laser. The kind and number of detectable substances is related to the spectral overlapping of the laser emission with the absorption bands of the trace gas molecules. Thus, the accessible wavelength range, tunability, and spectral resolution of the laser are of prime importance. With respect to minimum detectable concentrations (LPAS sensitivity), a laser with high output power P_L is a benefit, because the PA signal is proportional to P_L. The broad dynamic range is an inherent feature of LPAS and therefore is not affected by the choice of the radiation source. In contrast to remote-sensing methods, LPAS is a detection technique applied locally to samples enclosed in a PA cell. In order to still obtain some spatial resolution, either the samples have to be transported to the system, or the system has to be portable. The temporal resolution of LPAS is determined by the time needed for laser tuning and the gas exchange within the cell. Thus, a small volume PA cell and a fast tunable laser are a plus.

The availability of suitable laser sources plays a key role, as they control the sensitivity (laser power), selectivity (tuning range), and practicability (ease of use, size, cost, and reliability) that can be achieved with the photoacoustic technique. The CO_2 laser perfectly fits the bill for a trace gas monitoring system based on LPAS. This IR laser combines simple operation and high output powers. The frequency spacing between two adjacent CO_2-laser transitions range from 1 to 2 cm^{-1}. By contrast, the typical width of a molecular absorption line is approximately 0.05 to 0.1 cm^{-1} for atmospheric conditions. Since this is not a continuously tunable source, coincidences between laser transitions and trace gas absorption lines are mandatory. Fortunately, this does not hamper its applicability to trace gas detection, as numerous gases exhibit characteristic absorption bands within the wavelength range of the CO_2 laser which extends from 9 to 12 μm when different CO_2 isotopes are used. The CO_2 laser spectral output occurs in the wavelength region where a large number of compounds (including many industrial substances whose adverse health effects are a growing concern) possess strong characteristic absorption features and where absorptive interferences from water vapors, carbon dioxide, and other major atmospheric gaseous components may influence the measurements.

4

LPAS has emerged over the last decade as a very powerful investigation technique, capable of measuring trace gas concentrations at ppmV (parts per million by volume), or even sub-ppbV (parts per billion by volume) level. LPAS provide several unique advantages, notably the multicomponent capability, high sensitivity and selectivity, wide dynamic range, immunity to electromagnetic interferences, convenient real time data analysis, operational simplicity, relative portability, relatively low cost per unit, easy calibration, and generally no need for sample preparation. CO_2 LPAS offers a sensitive technique for detection and monitoring of trace gases at low concentrations and the spectroscopic system can be adaptable to a broad range of gases and vapors having absorption spectra in the IR with various applications in different disciplines, including: environmental pollutants monitoring (Schilt et al., 2004, Marinov and Sigrist, 2003), medical diagnostics (Bakhirkin et al., 2006, Narasimhan et al., 2001), life science field (Bijnen et al., 1996, Schilt et al., 2005), industrial process control (Besson et al., 2006), storage and transportation of fruit industry to control the ripening process of different fruits (Da Silva et al., 2001), detection of gases in hazardous work environment (Gondal et al., 2004), detection of various doping agents used by athlets (Fischer et al., 2006)) and many others (Dumitras et al., 2007)

The block diagram of the CO_2 laser photoacoustic spectrometer for gas studies, is shown in Fig. 1.1

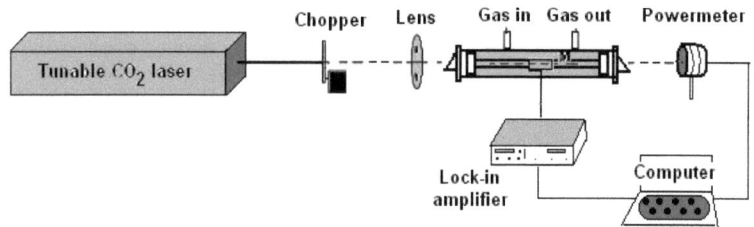

Figure 1.1. Block diagram of the CO_2 laser photoacoustic spectrometer.

The continuous wave, line-tunable, frequency-stabilized CO_2-laser beam is amplitude-modulated by a mechanical chopper operating at an acoustic resonance frequency of the PA cell chopped, focused by a ZnSe lens, and introduced in the PA

5

cell. After passing through the PA cell, the power of the laser beam is measured by a laser radiometer Rk-5700 from Laser Probe Inc. with a measuring head RkT-30. Its digital output is introduced in the data acquisition interface module together with the output from the lock-in amplifier. The gas concentration is proportional to the ratio between signal and laser power. All experimental data are processed and stored by a computer (Dumitras et al., 2007).

An advantage of PA spectroscopy as a tool for trace gas analysis is that very few photons are absorbed as the laser beam passes through the sample cell. As a result, notwithstanding the losses from absorption in the windows, the transmitted beam typically has sufficient power for analyzing samples in successive cells, via a multiplexing arrangement. A multiplexed PA sensor can be used to monitor many different samples simultaneously so that one instrument can be deployed to monitor up to 20 different locations within a clean room, industrial plant or other facility (Pushkarsky et al., 2002).

The term PA cell (or PA detector; both terms are used in the literature to describe the device in which the PA signal is generated and monitored) is reserved for the entire acoustic unit, including the resonator, acoustic baffles and filters, windows, gas inlets and outlets, and microphone(s). Finally, PA instrument (PA sensor) stands for a complete setup, including the PA cell, light source, gas handling system, and electronics used for signal processing.

It is interesting to mention that the *reverse* PA effect, called "sonoluminiscence", consists in the generation of optical radiation by acoustic waves, while the *inverse* PA effect is the generation of sound due to optical energy being lost from a sample, instead of being deposited in a sample as in the usual PA effect (Tam, 1986).

An extracavity arrangement was used because it has several advantages. In spite of a lower laser power available to excite the absorbing gas in the PA cell, a smaller coherent PA background signal makes it possible to increase the overall sensitivity of the instrument. Also, the dynamic range of the PA method is considerably reduced by intracavity operation. Optical saturation may occur for molecules with high absorption cross section while uncontrollable signal changes may be obtained at higher overall absorption in the PA cell, because the loss of light intensity influences the gain of the laser. This effect may cause erroneous results when the sample concentration changes are large.

Therefore, high-sensitivity single-and multipass extracavity PA detectors offer a simpler alternative to intracavity devices.

In this chapter, are described in detail the components of a sensor based on

LPAS principles. Special emphasis is laid on the home-built, frequency-stabilized, line-tunable CO_2-laser source and the resonant photoacoustic cell. Other aspects of a functional photoacoustic instrument, such as the gas handling system and data acquisition and processing, are outlined.

1.2. CO_2 lasers

In the author laboratory for gas studies it was designed, constructed and optimized a rugged sealed-off CO_2 laser (named LIR-25 SF), step-tunable on more than 60 vibrational-rotational lines and frequency stabilized by the use of plasma tube impedance variations detected as voltage fluctuations (the optovoltaic method) (Dumitras et al., 1981; Dumitras et al., 1985; Dutu et al., 1985). The glass tube has an inner diameter of 7 mm and a discharge length of 53 cm. At both ends of the tube were attached ZnSe windows at Brewster angle.

The laser is water cooled around the discharge tube. The dc discharge is driven by a high-voltage power supply. The end reflectors of the laser cavity are a piezoelectrically driven, partially (85%) reflecting ZnSe mirror at one end and a line-selecting grating (135 lines/mm, blazed at 10.6 μm) at the other. Piezoelectric ceramics such as lead zirconate titanate (PZT) can be used.

Figure 1.2. Homebuilt frequency stabilized CO_2 laser model LIR-25 SF.

The tunability of the CO_2 laser is presented in Figure 1.3. Can be notice the oscillation of 62 different vibrational-rotational lines in both the 10.4 μm and 9.4 μm bands. In this way, the laser was line tunable between 9.2 μm and 10.8 μm with powers varying between 1 and 6.5 W depending on the emitted laser transition. More than 20 lines had output powers in excess of 5 W.

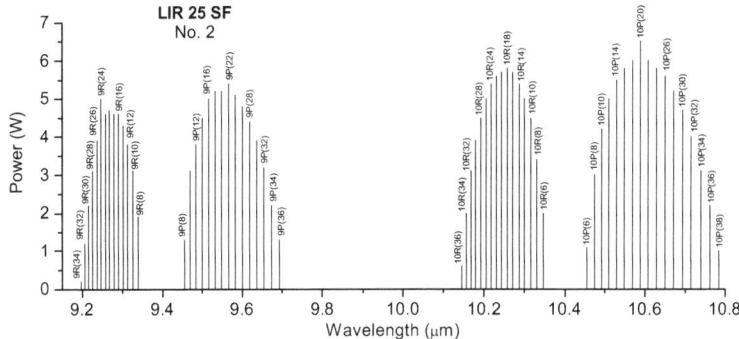

Figure 1.3. Tunability of the low power CO_2 laser with diffraction grating: P(max) = 6.5 W; Tunability: 9R(8) – 9R(34); 9P(8) – 9P(36); 10R(6) – 10R(36); 10P(6) – 10P(40); No. of lines: 62.

To investigate the possibility of using a high power laser in an extracavity configuration, was introduced in the experimental set-up a commercial CO_2 laser (Coherent GEM SELECT 50[TM] laser) (Figure 1.4) with output power till 50 W. When this laser is tuned on 10P(14) line, the maximum power delivered after chopper and focusing lens is 14.5 W.

Figure 1.4. Coherent GEM SELECT 50[TM] CO_2 laser in the experimental setup.

The tunability of high power CO_2 laser is presented in Figure 1.5. Can be notice the oscillation of 73 different vibrational-rotational lines between 9.2 μm and 10.8 μm with powers varying between 1 and 50 W depending on the emitted laser transition.

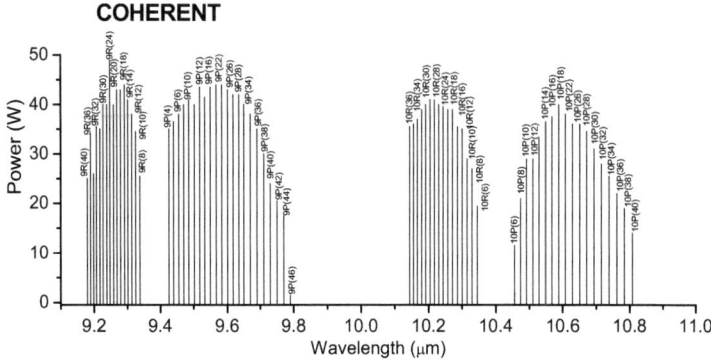

Figure 1.5. Tunability of the high power CO_2 laser with diffraction grating: P(max) = 50 W; Tunability: 9R(8) – 9R(40); 9P(4) – 9P(46); 10R(6) – 10R(36); 10P(6) – 10P(40); No. of lines: 73 (10P(14): λ = 10.53 μm; 9R(30): λ = 9.22 μm).

1.3 Mechanical chopper

The light beam was modulated with a high quality, low vibration noise and variable speed (4-4000 Hz) mechanical chopper model DigiRad C-980 or C-995 (30 slot aperture) operated at the appropriate resonant frequency of the cell (564 Hz). The laser beam diameter is typically 5 mm at the point of insertion of the chopper blade and is nearly equal to the width of the chopper aperture. An approximately square waveform was produced with a modulation depth of 100% and a duty cycle of 50% so that the average power measured by the powermeter at the exit of the PA cell is half the cw value. By enclosing the chopper wheel in a housing with a small hole (10 mm) allowing the laser beam to pass, chopper-induced sound vibrations in air that can be transmitted to the microphone detector as noise interference are reduced. A phase reference signal is provided for use with a lock-in amplifier (Dumitras et al., 2007, Dumitras et al., 2012b).

1.4 Lock-in amplifier

The generated acoustic waves are detected by microphones mounted in the cell wall, whose signal is fed to a lock-in amplifier locked to the modulation frequency. The lock-in amplifier is a highly flexible signal recovery and analysis instrument, as it is able to measure accurately a single-frequency signal obscured by noise sources many thousands of times larger than itself. It rejects random noise, transients, incoherent discrete frequency interference and harmonics of measurement frequency.

A lock-in measures an ac signal and produces a dc output proportional to the ac signal. Because the dc output level is usually greater than the ac input, a lock-in is termed an amplifier. The lock-in can also gauge the phase relationship of two signals at the same frequency. A demodulator, or phase-sensitive detector (PSD), is the basis for a lock-in amplifier. This circuit rectifies the signals coming in at the desired frequency. The PSD output is also a function of the phase angle between the input signal and the amplifier's internal reference signal generated by a phased-locked loop locked to an external reference (chopper).

A dual-phase, digital lock-in amplifier Stanford Research Systems model SR 830 was used with the following characteristics: full scale sensitivity, 2 nV - 1 V; input noise, 6 nV (rms)/$\sqrt{Hz}$ at 1 kHz; dynamic reserve, greater than 100 dB; frequency range, 1 mHz – 102 kHz; time constants, 10 μs – 30 s (reference > 200 Hz), or up to 30000 s (reference < 200 Hz).

The diverging IR laser beam is converged by a ZnSe focusing lens ($f = 400$ mm). In this way, a slightly focused laser beam is passed through the photoacoustic cell without wall interactions (Dumitras et al., 2007, Dumitras et al., 2012a, Dumitras et al., 2012b).

1.5 Photoacoustic cell

A PA cell was designed, constructed, and tested. An H-type cylindrical cell designed for resonant photoacoustic spectroscopy in gases is shown in Figure 1.6 The longitudinal resonant cell is a cylinder with microphones located at the loop position of the first longitudinal mode (the maximum pressure amplitude).

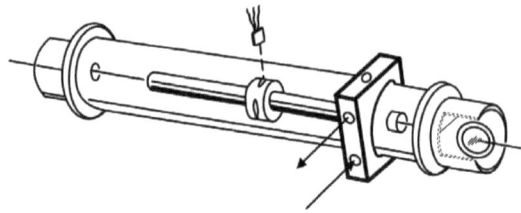

Figure 1.6. Schematic of the PA cell designed for the first longitudinal resonance mode.

The PA cell is made of stainless steel and Teflon to reduce the outgassing problems and consists of an acoustic resonator (pipe), windows, gas inlets and outlets, and microphones. It also contains an acoustic filter to suppress the flow and window noise. ZnSe windows at the Brewster angle are glued with epoxy (Torr-Seal) to their respective mounts.

The resonant conditions are obtained as longitudinal standing waves in an open tube (resonator) are placed coaxially inside a larger chamber. An open end tube type of resonator was used, excited in its first longitudinal mode. To achieve a larger signal, was chosen a long absorption path length ($L = 300$ mm) and an inner diameter of the pipe of $2r = 7$ mm. The fundamental longitudinal wave, therefore, has a nominal wavelength $\lambda_s = 2L = 600$ mm (and a resonance frequency $f_0 = 564$ Hz). The two buffer volumes placed near the Brewster windows have a length $L_{buf} = 75$ mm ($\lambda/8$) and a diameter $2r_{buf} = 57$ mm ($r_{buf} \cong 8r$).

The inner wall of the stainless steel resonator tube is polished. It is centered inside the outer stainless steel tube with Teflon spacers. A massive spacer is positioned at one end to prevent bypassing of gas in the flow system; the other is partially open to avoid the formation of closed volumes. Gas is admitted and exhausted through two ports located near the ends of the resonator tube. The perturbation of the acoustic resonator amplitude by the gas flow noise is thus minimized.

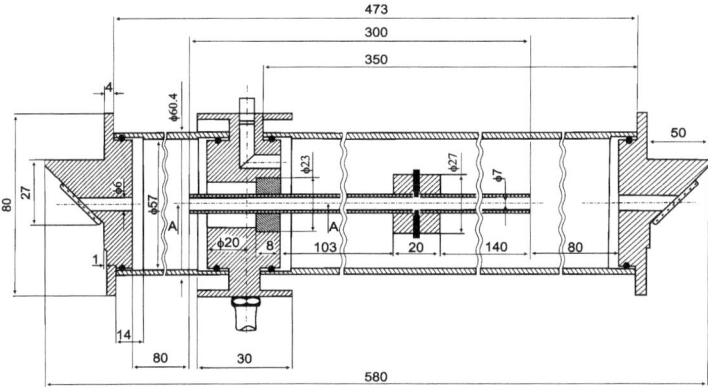

Figure 1.7. A resonant photoacoustic cell with buffer volumes.

The total cell volume is approximately 1.0 dm^3 (total length 450 mm, and inner diameter 57 mm, minus inner mechanical parts). For flowing conditions, however, it is advantageous to reduce the active volume of the cell. Especially if the flow rate is smaller than 1 L/h (16.6 sccm - standard cubic centimeters per minute), the replenish time for the 1.0 dm^3 cell becomes impractical. The buffer volume at the entrance port of the cell affects the renewal time τ considerably. The buffer volume is approximately 200 cm^3, yielding a time constant of 12 min ($\tau = V/R_{flow} = (200$ cm$^3)/(16.6$ cm^3/min) = 12 min). By reducing the buffer volume to 24 cm^3, with $r_{buf} \cong 3r$ (diameter 20 mm, length 75 mm), a τ of 1.5 min is obtained. However, an increased acoustical noise level was observed, due to the gas flow.

In photoacoustic measurements in the gas phase, microphones are usually employed as sensing elements of the acoustic waves generated by the heat deposition of the absorbing molecules. The most common microphones employed are miniature electret devices originally developed as hearing aids. The choice of a miniature microphone is particularly advantageous since it can be readily incorporated in the resonant cavity without significantly degrading the quality factor Q of the resonance. The frequency response of electret microphones extends beyond 10 kHz, and the response to incident pressure waves is linear over many orders of magnitude.

In the PA cells there are four Knowles electret EK-23024 miniature microphones in series (sensitivity 20 mV/Pa each at 564 Hz) mounted flush with the wall. They are situated at the loops of the standing wave pattern, at an angle of 90° to one another.

12

The microphones are coupled to the resonator by holes (1 mm diameter) positioned on the central perimeter of the resonator. The battery-powered microphones are mounted in a Teflon ring pulled over the resonator tube (Figure 1.8). It is of significant importance to prevent gas leakage from inside the resonator tube along the Teflon microphone holder, since minute spacing between the holder and resonator tube produces a dramatic decrease of the microphone signal and the Q value. The electrical output from these microphones is summed and the signal is selectively amplified by a two-phase lock-in amplifier tuned to the chopper frequency.

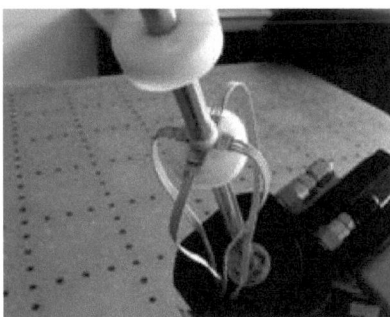

Figure 1.8. Teflon rings used to mount the microphones flush with the tube wall.

The resonance curve of our PA cell (cell response in rms volts) was recorded as a function of laser beam chopping frequency and the results are plotted in Figure 1.9. An accurate method is to construct the resonance curve point by point. In this case, the acoustic signal is measured at different fixed frequencies thus avoiding potential problems arising from the slow formation of a steady state standing wave in the resonator and the finite time resolution of the lock-in amplifier.

It is evident from these data that the cell resonance curve is fairly broad, implying that the absorption measurements would not be considerably affected if the frequency of the laser modulation or the cell resonance itself were to shift by a few hertz. PA cells exhibiting narrow resonances (as in the case of excitation of the radial modes) require tight control of both temperature and laser modulation frequency to avoid responsivity losses during the experiment.

The acoustic resonator is characterized by the quality factor Q, which is defined as the ratio of the resonance frequency to the frequency bandwidth between half power points. The amplitude of the microphone signal is $1/\sqrt{2}$ of the maximum amplitude at these points, because the energy of the standing wave is proportional to the square of the induced pressure. The quality factor was measured by filling the PA cell with 1 ppmV of ethylene buffered in nitrogen at a total pressure of 1 atm and by tuning the modulation frequency in 10 Hz increments (2 Hz increments near the top of the curve) across the resonance profile to estimate the half width, as described above.

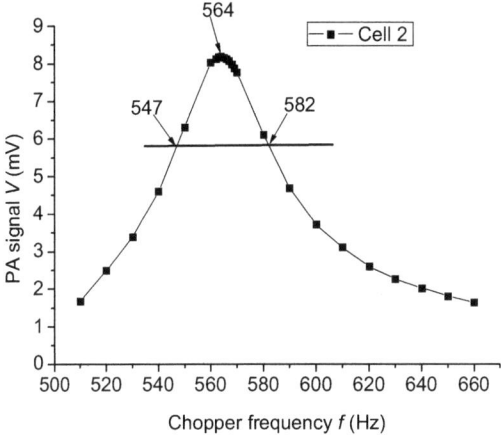

Figure 1.9. Resonance curve for the first longitudinal mode of the PA cell.

For this PA cell, the profile width at half intensity was 35 Hz, yielding a quality factor $Q = 16.1$ at a resonance frequency $f_0 = 564$ Hz. The experimentally determined resonance is not completely symmetric, as the curve rises steeply on one side and becomes less steep on the other side of the maximum. This asymmetry is caused by a coherent superposition of the standing acoustic waves in the detection region of the microphones (Karbach & Hess, 1985).

When the resonance contributions are included, the photoacoustic voltage signal can be obtained at a given operating frequency simply by multiplying the pressure response by the microphone responsivity ($V = pS_M$ and $\alpha_p = \alpha c$):

$$V = \alpha CS_M P_L c \,, \tag{1}$$

where: V (V) is the photoacoustic signal (peak-to-peak value); α (cm^{-1} atm^{-1}), the gas absorption coefficient at a given wavelength; C (Pa cm W^{-1}), the cell constant; S_M (V Pa^{-1}), the microphone responsivity; P_L (W), the cw laser power (unchopped value; 2x measured average value); and c (atm), the trace gas concentration (usually given in units of per cent, ppmV, ppbV or pptV).

This equation reveals that the photoacoustic signal is linearly dependent on laser power. Thus, sensitive measurements benefit from using as much laser power as is reasonably available. Moreover, the signal is directly dependent on the number of molecules in the optical path (trace gas concentration), which means that this technique is truly a "zero-baseline" approach, since no signal will be generated if the target molecules are not present (Dumitras et al., 2012a, Dumitras et al., 2012b).

The PA signal is linearly dependent on the absorption coefficient, cell constant, microphone responsivity, incident laser power, and absorbent trace gas concentration. Thus, by doubling Q (and consequently cell constant C), or the microphone responsivity, or the laser power, or the number of absorbing molecules in the optical path, the voltage will also double. The peak-to-peak value of the signal is obtained by multiplying by $2\sqrt{2}$ the rms voltage amplitude measured by the lock-in amplifier. As a rule, another parameter is used to characterize the PA cell, namely:

$$R = CS_M,\tag{2}$$

where R (V cm/W) is the (voltage) responsivity of the PA cell or the calibration constant. The cell constant C is multiplied by the responsivity of the microphone given in V/Pa units. A comparison of different PA cells can be made independently of the application in terms of this figure of merit. However, the cell characterization can be used only if a calibrated microphone is available. In this way, Eq. (27) becomes:

$$V = \alpha R P_L c.\tag{3}$$

To increase the detection sensitivity, we have to ensure: a) a cell constant as large as possible (optimization of the PA resonator); b) a large microphone responsivity; c) a laser power as high as possible, provided that saturation does not become a limiting factor; d) a narrow bandwidth of the lock-in amplifier, and e) a high absorption coefficient of the trace gas to be measured at the laser wavelength. It is also useful to increase the number n of microphones (connected in series), but this number is limited by the dimensions of the PA cell. The summation of the signals from the single microphones results in an n-times higher effective PA signal, because the total responsivity $S_{M\,tot}$ is increased n-fold, i.e.:

$$S_{M\,tot} = nS_M .$$ (4)

On the other hand, the incoherent noise only increases by $\sqrt{n}$. One thus obtains:

$$\text{SNR}_{tot} = \sqrt{n}\,\text{SNR} .$$ (5)

The minimum measurable voltage signal $V = V_{min}$ is obtained at SNR = 1, where the minimum detectable concentration $c = c_{min}$ can be recorded:

$$c_{min} = \frac{V_{min}}{\alpha P_L R} .$$ (6)

The sensitivity of PA instruments increases with the laser power, as $V \propto \alpha P_L$. However, the voltage signal does not depend on the length of the absorption path. Furthermore, in contrast to other techniques based on absorption spectroscopy, the response of the acoustic detector is independent of the electromagnetic radiation wavelength as long as the absorption coefficient is fixed. According to theoretical considerations, extremely low detection limits on the order of $\alpha_{min} = \alpha c_{min} \cong 10^{-10}$ cm^{-1} for 1 W incident laser power have been predicted (Sigrist, 1986) and experimentally proved (Harren et al., 1990). Such sensitivity makes it possible to detect many trace constituents in the sub-ppbV range.

1.6 Data acquisition and processing

The acquisition and processing of the recorded data was done with Keithley TestPoint software. TestPoint data acquisition software provides a development environment in which data acquisition applications can be generated. A graphical editor is provided for creating a user interface, or "panel", which the user sees and interacts with as the application executes. A user panel is made of pictorial elements that represent such things as switches, variable controls, numerical, text and selection boxes, bar displays, graphs, and strip charts. In addition, an application editor is provided, which ensures some interactive means of specifying how the visual elements on the user panel interact with the data sources and processing functions to achieve application goals. TestPoint uses an automated textual description of the operations carried out by each user panel element.

A modular software architecture was developed for controlling the experiments, collecting data, and preprocessing information. It helps automate the process of collecting and processing experimental results. The software controls the chopper frequency, transfers powermeter readings, normalizes data, and automatically stores files. It allows the user to set parameters such as the PA cell responsivity (a constant used to normalize raw data), gas absorption coefficient, number of averaged samples at every measurement point, sample acquisition rate, and total number of measurement points.

The software user interface allows the user to set or read input data and instantaneous values for the PA voltage (rms), average laser power after chopper, and trace gas concentration. Users may set experimental parameters for the PA cell responsivity and gas absorption coefficient. They are also provided with a text input to write a description of the experiments or take other notes. The user interface also provides data visualization (Fig. 1.10).

The software user interface contains three panels which display in real time the following parameters: CO_2 laser power level; PA signal; and trace gas concentration. Another window (countdown) indicates the number of remaining measurement points.

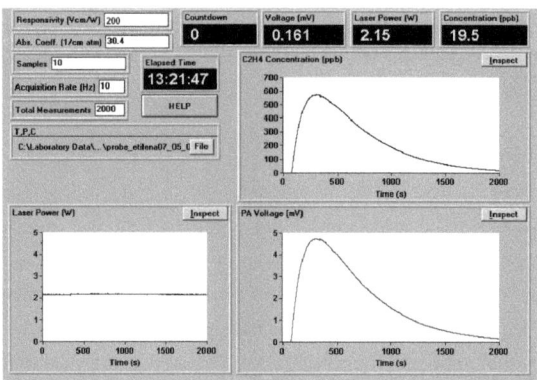

Figure 1.10. Software user interface used to record trace gas concentrations.

All settings and properties are stored to disk from session to session. In addition, a file may be automatically generated when running an experiment, including: a) **Laser power** stores powermeter readings of power incident on the sample as a function of time; b) **PA signal** stores the instantaneous values of the PA signal measured by the lock-in amplifier as a function of time; c) **Trace gas concentration** stores the time evolution of the trace gas concentration for a given laser wavelength (Dumitras et al., 2007, Dumitras et al., 2012a, Dumitras et al., 2012b).

1.7 Gas handling system

An important element in LPAS system is also the vacuum/gas handling system. The Teflon/stainless steel system can perform several functions without necessitating any disconnection. It can be used to ensure PA cell and gas purity, to pump out the cell, to introduce the sample gas in the PA cell at a controlled flow rate, and monitor the total and partial pressures of gas mixtures. (Dumitras et al., 2012a, Dumitras et al., 2012b).

A vacuum/gas handling system to be used in PA experiments was designed and implemented. The schematic of the gas handling chain is shown in Figure 1.11.

Gas transport lines throughout the gas mixing station were made of Teflon to minimize adsorption and contamination. The toggle valves V1-V17 and union tees T1-T11 were made of stainless steel. No valve grease was used. The PA cell gas inlet and outlet were connected to the gas handling system with Swagelok fittings. Connections to the inlet and outlet valves of the PA cell were made via flexible Teflon tubing so as to minimize the coupling of mechanical vibrations to the PA cell. The flexible lines also make it possible to position the PA cell during optical alignment.

The pressure of the gases added to the PA cell was determined by means of three Baratron pressure gauges (MKS Instruments, Inc.): 622A (0-1000 mbar), 122A (0-1000 mbar), and 122A (0-100 mbar), connected to a digital two-channel unit PDR-C-2C.

Thermal mass flowmeters, or mass flow controllers (MFC) were used, to deliver stable and known gas flows to the PA cell. The most critical processes will require flow measurement accuracies of 1% or better in the range 1000 to 10 sccm ($7x10^{-4}$ to $7x10^{-6}$ mol/sec; 1 sccm (at $0°C$) = $7.436x10^{-7}$ mol/sec). The digital MFCs sense the mass flow from the temperature difference between two temperature sensors in thermal contact with the gas stream and then process the information digitally with a microcontroller. The analog sensor output is amplified and digitized before it is sent to a microprocessor to compute the final control valve position. The gas flow in our gas handling system is adjusted by two gas flow controllers, MKS 1179A (0 - 1000 sccm) and MKS 2259CC (0 - 200 sccm), which are connected to a digital four-channel instrument MKS 247C.

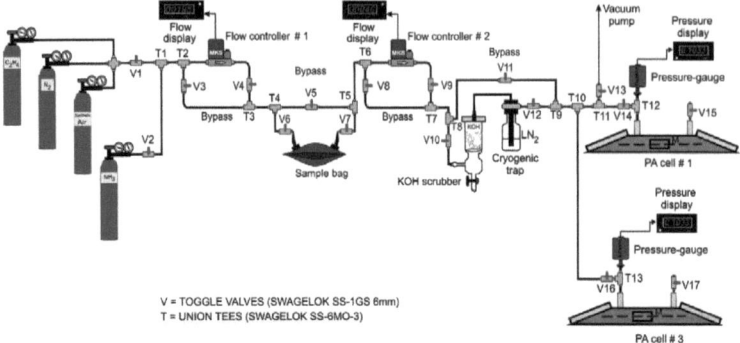

Figure 1.11. Gas handling system.

By using an adequate scrubber for CO_2 filtration, the CO_2 interference problem can be resolved. The CO_2 trap must neither alter the ethylene concentration level, nor introduce new interfering gases.

The flow rate was usually set at a low value of 30-600 sccm in all experiments in order to eliminate the acoustic noise of the gas flow, and all measurements were carried out with the PA cell at atmospheric pressure

As far as the sampling procedure is concerned, an extractive method was used, based on the collection of trace gas samples by some type of container or collecting medium and subsequent analysis in the laboratory (Dumitras et al., 2007, Dumitras et al., 2012a, Dumitras et al., 2012b)..

Conclusions

Gas sensing is of great interest in numerous areas. These include atmospheric studies (air pollutants, ozone detection, explosives), volcanic activity, agriculture (measurements of gaseous plant hormones), photochemistry (chemical processes monitoring, such as reaction rates, equilibrium constants, enthalpies or identification of different compounds, even isotopes, isomers and radicals), industrial process and workplace surveillance, gas certification, medical diagnosis, etc.

Trace gas sensing systems have to meet several requirements such as high detection sensitivity and selectivity, multicomponent capability, field suitability, etc. In this respect, devices based on tunable lasers combined with appropriate detection schemes are very attractive today.

The main components were lengthily described based on a general schematic. Special consideration were given to the frequency-stabilized, line-tunable CO_2-laser sources and the resonant photoacoustic cell. Other aspects of a functional photoacoustic instrument, such as the gas handling system and data acquisition and processing are outlined.

References

Bakhirkin Y. A., Kosterev A. A., Curl R. F., Tittel F. K Yarekha., D. A., Hvozdara L., Giovannini M., and Faist J. (2006). Sub-ppbv nitric oxide concentration measurements using cw thermoelectrically cooled quantum cascade laser-based integrated cavity output spectroscopy, *Appl. Phys. B*, Vol. 82, No. 1, pp. 149-154

Bell A.G. (1880). On the Production and Reproduction of Sound by Light. *Am. J. Sci.,* Vol.XX, pp. 305-324

Besson J. P., Schilt S., Sauser F., Rochat E., Hamel P., Sandoz F., Niklµes M., and Thevenaz L. (2006).Multi-hydrogenated compounds monitoring in optical fibre manufacturing process by photoacoustic spectroscopy, *Appl. Phys. B*, Vol. 85, pp. 343-348

Bicanic D. (Ed.) (1992). Photoacoustic and Photothermal Phenomena III, Springer Series in *Optical Sciences*, Vol.69, Springer, ISBN 978-0-387-55669-9, Berlin, Germany

Bijnen F. G. C., Harren F. J. M., Hackstein J. H. P., and Reuss J. (1996). Intracavity CO laser photoacoustic trace gas detection: cyclic CH4, H2O and CO2 emission by cockroaches and scarab beetles, *Appl. Opt,.* Vol. 35, No. 27, pp. 5357-5368

Da Silva M., Lima J., Sthel E., Marin E., Gatts C., Cardoso S., Campostrini E. , Pereira M. G., Campos A., Massunaga M., and Vargas H. (2001). Ethylene and CO_2 emission rates in tropical fruits investigated by infrared absorption techniques, *Anal. Sci.,* Vol. 17, pp. 534-537

Dumitras D.C., Dutu D.C., Comaniciu N., Draganescu V., Alexandrescu R., & Morjan I. (1981). Frequency Stabilized CO_2 Laser Design. *Rev. Roum. Phys.,* Vol.26, No.5, pp. 485-498, ISSN 1221-1451

Dumitras D.C., Dutu D.C., Draganescu V. and Comaniciu N. (1985). *Frequency Stabilization of CO_2 Lasers*. Preprint LOP-55, CIP Press, Bucharest, Romania

Dumitras D.C., Dutu D.C., Matei C., Magureanu A.M., Petrus M. and Popa C. (2007). Laser Photoacoustic Spectroscopy: Principles, Instrumentation, and Characterization. *J. Optoelectr. Adv. Mater.,* Vol.9, No.12, (December 2007), pp. 3655-3701, ISSN 1454-4164

Dumitras D. C., Bratu A. M. and Popa C., (2012a). CO_2 Laser Photoacoustic Spectroscopy: Principles, Intech, Croatia (Chapter I in *CO₂ Laser-Optimisation and Application*, ISBN 979-953-307-712-2, Ed. D. C. Dumitras .

Dumitras D. C., Bratu A. M. and Popa C., (2012b). CO_2 Laser Photoacoustic Spectroscopy: Instrumentation and Applications, Intech, Croatia; *Chapter II in CO_2 Laser-Optimisation and Application*, ISBN 979-953-307-712-2, Ed. D. C. Dumitras.

Dutu D.C., Draganescu V., Comaniciu N. and Dumitras D.C. (1985). Plasma Impedance and Optovoltaic Effect in Sealed-Off CO_2 Lasers. *Rev. Roum. Phys.,* Vol.30, No.2, pp. 127-130, ISSN 0035-4090

Fischer C., Bartlome R., and Sigrist M. W. (2006).The potential of mid- infrared photoacoustic spectroscopy for the detection of various doping agents used by athletes, *Appl. Phys. B,* Vol. 85, pp. 289-294

Gondal M. A., Dastageer A., and Shwehdi M. H. (2004).Photoacoustic spec- trometry for trace gas analysis and leak detection using di®erent cell geometries, *Talanta,* Vol. 62, pp. 131-141

Gusev V.E.and Karabutov A.A. (1993). *Laser Optoacoustics*, American Institute of Physics, ISBN 978-1-563-96036-9, Melville, NY, USA

Harren F.J.M. and Reuss J. (1997). Spectroscopy, Photoacoustic, In *Encyclopedia of Applied Physics*, Vol.19, G.L. Trigg (Ed.), pp. 413-435, VCH Publishers, ISBN 978-3-527-40478-0, New York, USA

Harren F.J.M., Bijnen F.G.C., Reuss J., Voesenek L.A.C.J. and Blom C.W.P.M. (1990). Sensitive Intracavity Photoacoustic Measurements with a CO_2 Waveguide Laser. *Appl. Phys. B,* Vol.50, No. 2, (February 1990), pp. 137-144, ISSN 0946-2171

Harren F.J.M., Cotti G., Oomens J. and te Lintel Hekkert S. (2000). Photoacoustic Spectroscopy in Trace Gas Monitoring, In *Encyclopedia of Analytical Chemistry*, Vol.3, R.A. Meyers (Ed.), pp. 2203-2226, Wiley, ISBN 978-0-471-97670-7 Chichester, UK

Hess P. (1983). Resonant Photoacoustic Spectroscopy, In *Topics in Current Chemistry*, Vol.111, F.L. Boschke (Ed.), pp. 1-32, Springer, ISBN 3-540-16403-0, Berlin, Germany

Hess P. (Ed.). (1989). *Photoacoustic, Photothermal and Photochemical Processes in Gases*, Topics in Current Physics, Vol.46, Springer, ISBN 978-0-387-51392-2, Berlin, Germany

Karbach A. and Hess P. (1985). High Precision Acoustic Spectroscopy by Laser Excitation of Resonant Modes. *J. Chem. Phys.,* Vol.83, No.3, (August 1985), pp. 1075-1084, ISSN 0021-9606

Mandelis A. (Ed.) (1992). *Progress in Photothermal and Photoacoustic Science and Technology*, Vol.1, Elsevier, ISBN 978-0-819-42450-1, New York, USA

Mandelis A. (Ed.) (1994). *Progress in Photothermal and Photoacoustic Science and Technology*, Vol.2, Prentice Hall, ISBN 978-0-131-47430-8, Englewood Cliffs, NJ, USA

Mandelis A. and Hess P. (Eds.) (1997). *Progress in Photothermal and Photoacoustic Science and Technology*, Vol.3, SPIE Press Book, ISBN 978-0-819-42450-1, Bellingham, WA, USA

Marinov D. and Sigrist M., (2003) Monitoring of road-traffic emissions with a mobile photoacoustic system., *Photochem. Photobiol. Sci.,* Vol. 2, No. 7, pp. 774-778

Miklós A., Hess P. and Bozoki Z. (2001). Application of Acoustic Resonators in Photoacoustic Trace Gas Analysis and Metrology. *Rev. Sci. Instrum.* Vol.72, No.4, (April 2001), pp. 1937-1955, ISSN 0034-6748

Miklós A., Lim C.-H., Hsiang W.-W., Liang G.-C., Kung A.H., Schmohl A. and Hess P. (2002). Photoacoustic Measurement of Methane Concentrations with a Compact Pulsed Optical Parametric Oscillator. *Appl. Opt.,* Vol.41, No.15, (May 2002), pp. 2985-2993, ISSN 0003-6935

Narasimhan L. R., Goodman W., and Patel C. K. N. (2001). Correlation of breath ammonia with blood urea nitrogen and creatinine during hemodialysis, *PNAS*, Vol. 98. No. 8, pp. 4617-4621

Pao Y.-H. (Ed.) (1977). *Optoacoustic Spectroscopy and Detection*, Academic, ISBN 978-0-125-44150-6, New York, USA

Patel C.K.N. and Tam A.C. (1981). Pulsed Optoacoustic Spectroscopy of Condensed Matter. *Rev. Mod. Phys.,* Vol.53, No.3, (July-September 1981), pp. 517-554, ISSN 0034-6861

Pushkarsky M.B., Weber M.E., Baghdassarian O., Narasimhan L.R. and Patel C.K.N. (2002). Laser-Based Photoacoustic Ammonia Sensors for Industrial Applications. *Appl. Phys. B,* Vol.75, No.4-5, (April 2002), pp. 391-396, ISSN 0946-2171

Rosencwaig A. (1980). *Photoacoustics and Photoacoustic Spectroscopy*, Chemical Analysis Vol.57, Wiley, ISBN 978-0-471-04495-4, New York, USA

Schilt S., Besson J. P., Thevenaz L., and Gyger M., (2005).Continuous and simultaneous multigas monitoring using a highly sensitive and selective photoacoustic sensor, in *CLEO/QELS and PhAST*

Schilt S., Thevenaz L., Nikles M., Emmenegger L., and Huglin C. (2004). Ammonia monitoring at trace level using photoacoustic spectroscopy in industrial and environmental applications. *Spectrochim. Acta, Part A*, Vol. 60, No. 14, pp. 3259-3268

Schmid T. (2006). Photoacoustic Spectroscopy for Process Analysis. *Anal. Bioanal. Chem.,* Vol.384, No.5, (March 2006), pp. 1071-1086, ISSN 1618-2642

Sigrist M.W. (1986). Laser Generation of Acoustic Waves in Liquids and Gases. *J. Appl. Phys.,* Vol.60, No.7, (October 1986), pp. R83-R121, ISSN 0021-8979

Sigrist M.W. (Ed.) (1994). *Air Monitoring by Spectroscopic Techniques*, Vol.127, Wiley Chemical Analysis Series, Wiley, ISBN 978-0-471-55875-3, New York, USA

Tam A.C. (1986). Applications of Photoacoustic Sensing Techniques. *Rev. Mod. Phys.,* Vol.58, No.2, (April-June 1986), pp. 381-431, ISSN 0034-6861

West G.A. (1983). Photoacoustic Spectroscopy. *Rev. Sci. Instrum.,* Vol.54, No.7, (July 1983), pp. 797-817, ISSN 0034-6748

Zharov V.P. and Letokhov V.S. (1986). *Laser Optoacoustic Spectroscopy*, Vol.37, Springer, ISBN 978-3-540-11795-4, Berlin, Germany

Chapter 2.
Calibration and detection limit

2.1 Introduction

Trace gas monitoring requires extreme performances in terms of sensitivity and selectivity (Besson, 2006). CO$_2$ lasers present the main advantage of their high power and makes thus possible to reach extreme sensitivities in the ppbV level. This IR laser combines simple operation and high output powers.

This chapter presents an extremely sensitive apparatus based upon LPAS methods which can be used for the detection and measurement of trace gases at very low concentrations. It was designed and characterized two experimental set-ups with the PA cell in an external configuration: the first one with a low power CO$_2$ laser where the saturation effects are negligible, and a second one with a high power CO$_2$ laser where the saturation effects are important and have to be taken into consideration. All measurements were done in nitrogen and ethylene with the 10P(14) line of a continuous wave (cw) CO$_2$ laser. It was obtained a minimum detectable concentration better by more than a factor of 10 compared to the best results previously reported in the literature.

A precise measurement of the absorption coefficients of ethylene at CO$_2$ laser wavelengths are presented. Noises and other limiting factors which determine the ultimate detection sensitivity are measured. The interfering gases play an important role both in limiting the sensitivity of the method and in the multicomponent analysis of the atmosphere and experimental measures for reducing the influence of interfering gases in a single component measurement and the methods used in multicomponent analysis are discussed

2.2 Low power vs. high power lasers

A PA signal may become saturated due to either a large concentration of the measured analyte or high laser power levels. By increasing laser intensity, the excitation pumping rate of the molecules grows higher, and a molecule is more likely to absorb a nearby photon before it relaxes to the ground state. So, as the molecules in the excited state increase in numbers, the number of molecules which can absorb laser radiation is reduced. The gas actually becomes as though more transparent to laser radiation, and the effective absorption coefficient per unit laser power is lowered; this is called laser power saturation. Saturation due to nonlinear absorption of the laser power only occurs in focused high-power laser beams or when the PA cell is placed intracavity in a laser, so that the laser power can be on the order of tens of watts or even higher than 100 W. The pumping rate to a higher vibrational-rotational level is proportional to the laser light intensity; in the case of saturation it exceeds the collisional de-excitation rates.

Two experimental set-ups with the PA cell in an external configuration were designed and characterized: the first one with a low power CO_2 laser where the saturation effects are negligible, and a second one with a high power CO_2 laser where the saturation effects are important and have to be taken into consideration. All measurements were done in nitrogen and ethylene with the 10P(14) line of a cw CO_2 laser.

To characterize the operation of the laser PA system at low laser power where saturation effects can be neglected, was used a homebuilt, line-tunable and frequency stabilized CO_2 laser. This laser has a maximum power of 6.5 W on 10P(20) line and a tunability on 62 vibrational-rotational lines in all four spectral bands. In ethylene (9.88 ppmV diluted in pure nitrogen), the measurements were made on 10P(14) line, where the absorption coefficient is maximum (30.4 $cm^{-1}atm^{-1}$). The laser power through PA cell (after chopper and focusing lens) is 2.2 W. It was noticed that by varying the laser power inside the cell in the range 0.5-2.2 W, no saturation effect was evident. The resonance frequency corresponds to the resonant excitation of the first longitudinal mode of the PA cell (depends on its length). The acoustic resonator is characterized by the quality factor Q, which is defined as the ratio of the resonance frequency to the frequency bandwidth between half power points. The amplitude of the microphone signal is $1/\sqrt{2}$ of the maximum amplitude at these points, because the energy of the standing wave is proportional to the square of the induced pressure. For this PA cell, the profile width at half intensity was 35 Hz, yielding a quality factor Q

= 16.1 at a resonance frequency f_0 = 564 Hz. The quality factor has a typical value for the resonant excitation of the first longitudinal mode (Q = 15-50 (Harren et al.,1997, Pushkarsky et al., 2002, Harren et al., 1990b, Bijnen et al., 1996)) or even Q = 70 for a multipass resonant PA cell (Nägele et al., 2000).

The value of the cell responsivity (R = 280 V cm/W) is one of the highest obtained till now. Most published papers give values between 37 and 200 V cm/W (Harren et al., 1990a, Harren et al., 1990b, Ryan et al., 1983, Rooth et al., 1990, Fink et al., 1996) and only few reports obtained similar values (Harren et al., 1997, Nägele et al., 2000) or a higher value (Bijnen et al., 1996) (by using a microphone with a very large responsivity).

To investigate the possibility of using a high power laser in an extracavity configuration, was introduced in the experimental set-up a commercial CO_2 laser (Coherent GEM SELECT 50^{TM} laser) with output power till 50 W and tunable on 73 different lines (Dumitras et al., 2010). When this laser is tuned on 10P(14) line, the maximum power delivered after chopper and focusing lens is 14.5 W.

By changing the laser power inside the PA cell was attempted either to modify the input power in the laser (RF power supply) or to introduce a beam splitter in the path of the laser beam. Unfortunately, both methods change significantly the beam path inside the PA cell, thus perturbing unacceptably the results of the measurements. The waveguide laser has a poor beam pointing because it has a short optical resonator and the variation of the transverse RF excitation modifies the laser gain profile.

The saturation effects at high laser power were investigated by using the method of truncation of a gaussian laser beam. This approach was possible because the laser beam is very close to a gaussian beam (M^2 < 1.1).

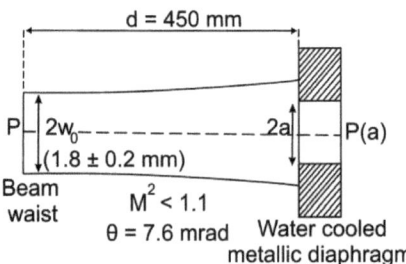

Figure 2.1. Attenuation of a laser beam by a diaphragm.

The method consists in passing the beam through an aperture with known diameter (Figure 2.1). To avoid deformations owing to heating, was used water cooled metallic diaphragms with diameters between 1.42 mm and 5.03 mm. All diaphragms were placed at a distance of 450 mm from the beam waist of the laser.

When a gaussian beam of radius w is truncated by an aperture of radius a (Figure 2.2), the power transmitted through the aperture is $T = P(a)/P = 1 - \exp(-2a^2/w^2)$. When $2a = 2w$, $T \cong 86\%$, that is 86% of the laser power is transmitted through the aperture (this is known as 86% criterion). When $2a = \pi w$, $T \cong 99\%$, that is 99% of the laser power is transmitted through the aperture resulting 99% criterion. This formula offers a possibility to measure precisely the diameter of the laser beam at the position of the diaphragm. By knowing the radius of the aperture (a) and by measuring the laser power before and after the aperture (P and $P(a)$, respectively), it can be determined immediately the radius of the laser beam (w). As it can be seen in Figure 2.3, by using five different diaphragms, the resulting average diameter is $2w = (7.09 \pm 0.2)$ mm, with an error of less than 3%.

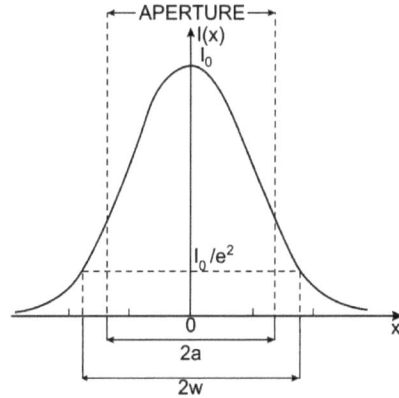

Figure 2.2. Truncation of a gaussian laser beam.

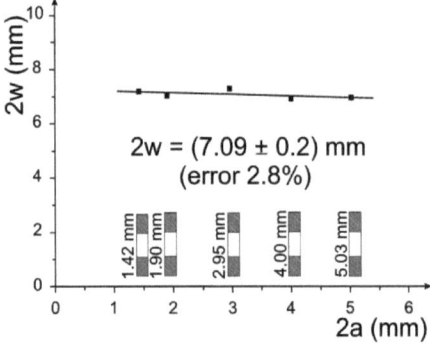

Figure 2.3. Measurement of laser beam diameter by the method of truncation.

Figure 2.4 shows the attenuation of the laser beam when different diaphragms were placed in its path. The solid line is the theoretical curve given by the above equation. By introducing these five diaphragms, the laser power was varied between less than 2 W and near 10 W. In this way, can be investigate the laser power range from low power where saturation effects have no significance till high power where saturation effects manifest strongly.

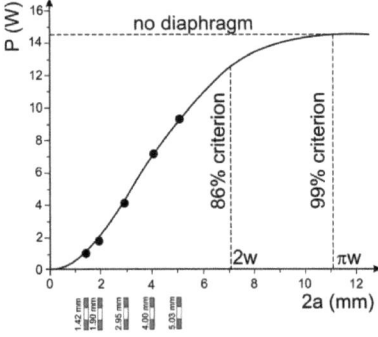

Figure 2.4. Variation of laser power function on diaphragm aperture size.

The influence of saturation was investigated by measuring the dependence of the PA cell responsivity function on laser power (Figure 2.5). From low laser power regime (under 2 W) where the saturation effects are not important till high power regime (14.5 W, no diaphragm), the PA cell responsivity decreases from 312 V cm/W till 52 V cm/W, that is by a factor of 6. It can be observed that the saturation effects manifest immediately as the laser power is increased more than 2 W. The cause of the saturation is that the collisional relaxation to de-excite the molecules cannot keep up with the excitation rate by the laser beam intensity. By increasing laser intensity, the excitation pumping rate of the molecules grows higher and a molecule is more likely to absorb a nearby photon before it relaxes to the ground state. So, as the number of molecules in the excited state increases, the number of molecules which can absorb laser radiation is reduced. That is why was introduced a supplementary scale in Figure 2.5, representing the cell responsivity function on the intensity of the focused beam inside the PA cell. A thorough analysis of the laser beam propagation through the focusing lens and in the PA cell was done (Dumitras et al., 2007). For a lens with a focal length of 400 mm, results in this case a beam diameter at beam waist of 0.89 mm. It can be remarked that saturation starts at laser intensities greater than 0.5 kW/cm². This result shows that saturation effects manifest even at low laser intensities. Previously, Harren et al. (Harren et al., 1990a) observed a strong saturation at a much higher intensity (200 kW/cm²; the laser power was ten times higher and the beam area was ten times smaller than in our case at 13 W), when the PA cell was placed intracavity of a waveguide CO₂ laser. In conclusion, high power lasers could be used in PA systems, but saturation effects should be taken into consideration (by making a correlation between the PA cell responsivity and the working laser intensity, as in Figure 2.5).

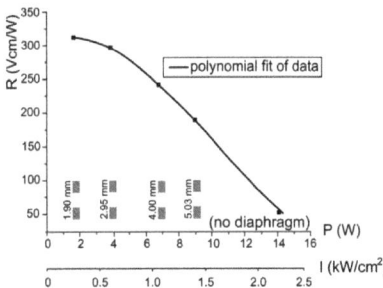

Figure 2.5 Saturation effects measured as the dependence of the PA cell responsivity function on laser power and on laser intensity in the focal spot.

An important question is: what happens with the system noises when a diaphragm is introduced in the laser beam path? We proceeded to record the coherent PA background signal on laser power (with and without a diaphragm) and the results are given in Figure 2.6. The background signal is huge when a diaphragm is inserted into the system, being of more than 50 times higher than in the case that no diaphragm limits the laser beam. Truncation distorts the intensity pattern of the transmitted beam in both the near-field (Fresnel) and far-field (Fraunhofer) regions.

The diffraction effects on an ideal gaussian beam of a sharp-edged circular aperture even as large as $2a = 2w$ (99% criterion) will cause near-field diffraction ripples with an intensity variation $\Delta I/I \cong \pm 17\%$ in the near field, along with a peak intensity reduction of $\cong 17\%$ on axis in the far field (Siegman, 1986). In conclusion, the method of diaphragms used to measure the saturation effects is applicable, but in a laser PA system used in practice an aperture has never to be introduced.

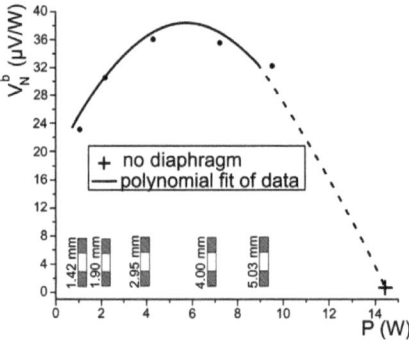

Fig. 2.6. Dependence of coherent PA background signal on laser power (with and without a diaphragm).

A comparison of low laser power vs. high laser power configurations is presented in Table 4. It seems reasonable that high power lasers could be used in PA instruments provided that the saturation is considered and compensated.

Table 2. Comparison of low vs. high laser power configurations

Parameter	Low power	High power	Factor
Output laser power (W)	5.5	33	> 6.0
Average laser power (at cell exit) (W)	2.2	14.5	> 6.6
Coherent photoacoustic background signal (μV/W)*	2.7	0.7	< 4.0
Cell responsivity R (V cm/W)**	280	312	> 1.1
Signal saturation**	Small	Very high	> 6.0
Minimum detectable concentration c_{min} (ppbV)**	0.9	0.21	< 4.3
c_{min} (ppbV)***	0.04	0.009	
Minimum detectable absorptivity α_{min} (cm^{-1})**	2.7×10^{-8}	0.64×10^{-8}	< 4.3
Best value previously reported (Harren et al., 1990) (c_{min} = 3.8 ppbV)**	Better 4.2 x	Better 18 x	

10P(14) laser line in N$_2$* ($\alpha = 0$ cm^{-1}atm^{-1}), C$_2$H$_4$** ($\alpha = 30.4$ cm^{-1}atm^{-1}), and SF$_6$*** ($\alpha = 686$ cm^{-1}atm^{-1}).

To this date, the minimum detectable concentrations obtained in ethylene (0.9 ppbV with a low power laser and 0.21 ppbV with a high power laser) are the best values reported in the literature.

2.3 Measurement of ethylene (C$_2$H$_4$) absorption coefficients

Photoacoustics is emerging as a standard technique for measuring extremely low absorptions independent of the path length and offers a degree of parameter control that cannot be attained by other methods. Radiation absorption by the gas creates a pressure signal which is sensed by the microphone. The resulting signal, processed by a phase sensitive detector, is directly proportional to the absorption coefficient and laser power (or laser power absorbed per unit volume). The sensitivity of the technique is such that absorptions of $<10^{-7}$ cm^{-1} can be measured over path lengths of a few tens of centimeters. The small volume of the chamber makes it possible to accurately control the gas parameters, and the system can be operated with static fills or in continuous gas flow mode.

The set of values of the absorption coefficients α, for all laser wavelengths, for a particular gas or vapor and at a common concentration is called the optoacoustic absorption spectrum or signature and is unique to a combination of vapor and laser. These signatures or "fingerprints" are absolute entities, unique only to the laser frequency and species, which provide the specifics of instrument performance in terms of detection limit and interference rejection (Cristescu et al., 2000b).

To improve the measurement of ethylene absorption coefficients, a special procedure was followed (Dumitras et al., 2012). Prior to each run, the gas mixture was flowed at 100 sccm for several minutes to stabilize the boundary layer on the cell walls, since a certain amount of adsorption would occur and possibly influence background signals; after this conditioning period, the cell was closed off and used in measurement. For every gas fill with 0.96 ppmV ethylene buffered in pure nitrogen, the responsivity of the cell was determined supposing an absorption coefficient of 30.4 cm^{-1}atm^{-1} at 10P(14) laser transition. After measurements at all laser lines, the cell responsivity was checked again, to eliminate any possibility of gas desorption during the measurement. The partial pressure of ethylene was enough to have significant PA signals for all laser lines and low enough to be far away from the saturation regime (observations were only made at a C_2H_4 concentration of 100 ppmV). The α values at each laser line were obtained from Eq. 1 using the measured PA signal and laser power (the cell responsivity and ethylene concentration were known). An average over several independent measurements at each line was used to improve the overall accuracy of the results. The values to be presented are thought to be the best published to date.

The absolute magnitudes of the absorption coefficients were calculated as mean values of several independent measurements. An absorption coefficient corresponding to each CO$_2$ laser transition was determined from two sets of 50 different measurements. Every set of measurements was initiated by the frequency stabilization of a given line of the CO$_2$ laser. From one set of measurements to

31

another, the closed loop of the frequency stabilization circuit was interrupted, the laser was tuned again to the top of the gain curve, and then the frequency stabilization was set and checked by watching the long term stability. Inside one set, 50 independent measurements were made at a rate of one per second to assess reproducibility. From one measurement to the next, the error measurement of the absorption coefficient was calculated as the ratio between the maximum difference (maximum value minus minimum value) and the average value. The final value of the ethylene absorption coefficient is given by the arithmetic mean of the two sets of measurements, while the absorption coefficient error is chosen as the larger value of the two sets. The same procedure was applied for every absorption coefficient of ethylene.

To measure the absorption coefficients of ethylene, the software user interface was modified to allow that the laser power, PA signal, and calculated absorption coefficients function on time (or number of measurements) be recorded on different panels.

The results of measurements for ethylene are given in Figure 2.7.

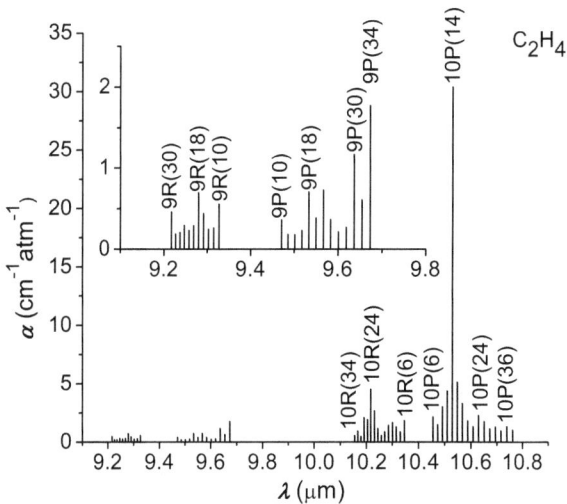

Figure 2.7. Absorption coefficients of ethylene at CO_2 laser wavelengths. The inset shows an enlarged view of the measurements for the 9-μm band.

The result is a spectral representation unique to that molecule. As a consequence of the superposition of different pressure-broadened C_2H_4 transitions (ν_7 vibration), a strong absorption is obtained at the 10P(14) laser line (absorption coefficient of 30.4 $cm^{-1}atm^{-1}$ at 949.479 cm^{-1}). C_2H_4 has weaker absorption coefficients at the 10P(12) and 10P(16) CO_2 laser transitions (4.36 $cm^{-1}atm^{-1}$ at 951.192 cm^{-1} and 5.10 $cm^{-1}atm^{-1}$ at 947.742 cm^{-1}, respectively). Also, in Figure 2.7 ethylene is seen to possess moderately strong absorption profiles within the 9.4-μm band.

The present work was carried out using a methodology which gave the best possible control over the ethylene partial pressure and background signals. The background levels and calibration of the PA cell were checked before and after every experimental run. The present study is considered reliable, particularly in view of the careful attention that was paid to controlling the gas composition and noise signals. No apparent fault could be found with either the apparatus or methodology that would account for the discrepancy by factors of 2-5 from other reported data in the case of the CO_2 laser 9R lines.

2.4 Noises in photoacoustic measurements

Noise plays an important role in all photoacoustic measurements and is of particular importance in the detection of ultralow gas concentrations, because the noise level limits the ultimate sensitivity. In the photoacoustic literature, the detection level is usually defined by the signal-to-noise ratio, where the noise is given by the microphone signal measured with the laser light blocked. However, when light hits the PA cell, an additional background signal is generated which exists even when the absorbing species are not present in the detector. The background signal is often larger than the noise signal, and therefore the detection limit or sensitivity has to be defined by the signal-to-background ratio (SBR) in most experiments. Unfortunately, it is common practice to consider only the SNR. This procedure yields an extrapolated detection limit that may be far too small. The background signal is usually determined with a nonabsorbing gas, such as nitrogen, in the PA detector. It is influenced by many system properties, such as the pointing stability, the beam divergence, and the diameter of the laser beam.

For photoacoustic spectroscopy, "noise" often has a structure that is coherent with the signal from the target species, and therefore should more appropriately be treated as a background signal, not as noise. The background signal can be determined by measuring the acoustic signal in the absence of absorbers (i.e., with pure nitrogen), but with the same flow and in the same pressure conditions as those used for the sample gases.

The sensitivity-limiting factors which are encountered in LPAS can be classified into three categories:

a) Electrical noise, meaning any random fluctuation, whether electronic or acoustic, which does not have a fixed phase relation with the modulation of the laser intensity. It determines the ultimate detector sensitivity.

b) Coherent acoustic background noise, meaning a signal caused by the modulation process, but not attributable to the presence of the light beam in the PA cell. This signal is at the same frequency as, and locked in phase with respect to, the laser intensity modulation.

c) Coherent photoacoustic background signal. This signal, which is always present in the PA detector, is caused by the laser beam, yet not by light absorption in the bulk of the gas. Rather it is due to laser beam heating of the windows and of the absorbates at their surfaces, and heating of the PA resonator walls by the reflected or scattered light owing to imperfections of the focusing lens, windows and inner walls of the PA resonator. This signal is in phase with, and at the same frequency as, the laser intensity modulation. Therefore, it is not filtered out by the lock-in amplifier connected to the microphone. Thus, a background signal proportional to the laser power becomes the main factor that limits sensitivity.

The background signal in the PA cells may arise from several sources, some of which are listed below (Gerlach and Amer, 1980):

(1)   Window surface absorption: the molecules absorbed on the window surface and/or the window surface itself absorb the modulated laser radiation, and the resulting gas heating in the cell generates a pressure pulse.

(2)   Window bulk absorption: even the highest quality ZnSe window substrates exhibit a residual window absorption of $\sim 10^{-3}$ cm^{-1}.

(3)   Off-axis radiation within the cell: light scattered from the windows and at the edge of the chopper blade may strike the inside walls of the PA resonator, where it may be absorbed and produce a signal.

(4)   Light scattering or absorption due to microaerosols.

(5)   Small amounts of contamination that may outgas from the cell materials, seals, and so forth.

The detection limit of the PA cell is determined by the combined effect of the intrinsic stochastic noise of the microphone, acoustic background noise, and photoacoustic background signal. Background signals are deterministic, and to the extent that they can be quantified and minimized, do not reduce the performance of the cell significantly. The detection limit is defined either at a signal-to-noise ratio of unity (SNR = 1) or at a signal-to-background ratio of unity (SBR = 1).

The amplifier input noise and microphone noise are gaussian in nature, that is, the amount of noise is proportional to the square root of the bandwidth in which the noise is measured. All of these noise sources are incoherent. The input noise of the SR830 lock-in amplifier used in these experiments is about 6 nV (rms)/$\sqrt{\text{Hz}}$. Microphone noise, which is manifested as a noise voltage present at the microphone output terminals, can be expressed as a product between the normalized noise pressure value owing to both thermal agitation of the diaphragm and cartridge responsivity at the corresponding frequency and the square root of the measurement bandwidth. The electrical noise of Knowles EK models electret microphones is 40 nV (rms)/$\sqrt{\text{Hz}}$. The overall random noise of multiple sources is determined by taking the square root of the sum of the squares of all the individual incoherent noise figures. For gaussian noise, the peak-to-peak value is about 5 times the rms noise value, while for the two other types of noises, the rms value must be multiplied by a factor of $2\sqrt{2} \cong 2.8$ to obtain the peak-to-peak amplitude. Electrical noise usually has a broadband frequency spectrum and can be reduced efficiently by narrowband filtering of the signal, as is done in the phase sensitive detection. A detection bandwidth of 0.25 Hz was set (a time constant of 1 second) in all measurements. Electrical noise can be reduced by using state-of-the-art (and therefore very expensive) lock-in amplifiers and/or by using longer time averaging (the noise decreases with the square root of the averaging time) at the cost of longer measurement times.

The two types of coherent background, however, are extremely narrowband signals at the same frequency as the modulation and hence cannot be filtered out. In addition, since the signal and the coherent photoacoustic background signal are both proportional to laser power, no improvement will be achieved as the laser power is increased.

The limiting electrical noise measured at resonance frequency was $v_N^e = 0.15\ \mu V$ /$\sqrt{\text{Hz}}$. At atmospheric pressure, the acoustic background noise was $v_N^{ac} = 3.1\ \mu V$ (at resonance frequency) under normal working conditions. A photoacoustic background signal of v_N^b (nitrogen) = 2.3 $\mu V/W$ was observed, in phase and at resonance frequency, as the cell was filled with pure N_2 at atmospheric pressure.

When the sample gas is flown continuously through the detector, acoustical noise can be produced, if the gas flow is turbulent, if acoustical noise from the surroundings is coupled directly into the detector sample space or into the tubes connected to the detector and then propagated into the detector, or if acoustic disturbances from the pump running the sample gas through the detector are propagated through the tubes. Thick detector and tube walls, small flow rates, mounting of the cell and chopper in separate sound insulating boxes, etc. must be chosen to suppress these noise contributions.

The background signal can be minimized by placing the windows at nodes of the mode being excited and by introducing buffer volumes at both ends of the cell. The ratio of buffer to resonator diameters must be large enough, and the buffer length has to be equal to one-fourth of resonator length.

2.5 Removal of interfering gases in breath biomarker measurements

Interference of other absorbing substances may impair the theoretical detection limit in a multicomponent analysis of the real samples. Such interference may be caused by other molecular systems present in the environment or substances that are entrained by the carrier flux. If an interfering species is present in the environment, its effect can be minimized by either the introduction of scrubbers and cryogenic traps or the use of dual beam techniques using two PA cells.

The CO_2 laser spectral outputs occur in the wavelength region where a large number of compounds possess strong absorption features and where absorptive interferences from water vapors, carbon dioxide, and other major atmospheric gaseous components may influence the measurements.

The breath air is a mixture of nitrogen, oxygen, carbon dioxide, water, inert gases, and traces of volatile organic compounds (VOCs) (Table 3). The matrix elements in breath air vary widely from person to person, both qualitatively and quantitatively, particularly for VOCs. More than 1000 trace VOCs have been distinguished in human breath air, at concentrations from ppmV to pptV levels (Ca and Duan, 2006, Wang and Sahay, 2009). Only a small number of VOCs are common to everyone, including isoprene, acetone, ethane, and methanol, which are products of core metabolic processes. In addition to these VOCs, exhaled NO, H_2, NH_3, and CO are related to health condition and can reflect a potential disease of the individual or a recent exposure to a drug or an environmental pollutant.

Table 3. Concentration of different components in inhaled and exhaled air

Component	Inhaled air (%)	Exhaled air (%)
Nitrogen	78.0	78.0
Oxygen	21.0	16.0
Carbon dioxide	0.04	3.0-5.0
Argon	0.93	1.0
Water	2.0	5.0-6.0
Other	0.01	
ammonia		250×10^{-9} (250 ppb)
ethylene		6×10^{-9} (6 ppb)

A healthy adult human has a respiratory rate of 12-15 breaths/min at rest, inspiring and expiring 6-8 L of air per minute. O_2 enters the blood and CO_2 is eliminated through the alveoli. When the end-tidal concentration of CO_2 in healthy persons is measured, a large change of CO_2 concentration is observed between the inhaled air ($\sim$ 0.04%) and the exhaled air ($\sim$ 4%) (Folke et al., 2003). The exact amount of exhaled CO_2 varies according to the fitness, energy expenditure and diet of a particular person, with regular values of 3-5%. Due to this high concentration of carbon dioxide in the breath and because CO_2 laser lines are absorbed by this gas, it is necessary to remove most of the carbon dioxide from the exhaled air by introducing a scrubber filled with a chemical active agent, like (Bratu et al., 2011).

Due to the exact coincidence of the CO_2 vibrational-rotational transitions with the CO_2 laser lines, carbon dioxide at high concentration in comparison with trace gases like C_2H_4 is inevitably excited by CO_2 laser radiation and the related photoacoustic signal may exceed the trace signal by many orders of magnitude. The absorption coefficient increases strongly with temperature, but it is independent of the CO_2 concentration over a wide range. Ethylene can be excited by the 10P(14) line of the CO_2 laser, where the maximum absorption coefficient $\alpha(C_2H_4)$ has a value of 30.4 cm^{-1} atm^{-1} and ammonia by the 9R(30) line where $\alpha(NH_3)$ = 56 cm^{-1} atm^{-1} (Dumitras et al., 2011). A 4% concentration of CO_2 has an absorption strength comparable to 2760 ppbV of C_2H_4 (at the 10P(14) laser line, $\alpha(CO_2)$ = 2.1x10^{-3} atm^{-1}cm^{-1} and $c(C_2H_4) = c(CO_2)\alpha(CO_2)/\alpha(C_2H_4)$. This equivalent ethylene concentration was found also experimentally (see Figure 2.9, measurement without trap). So, the photoacoustic signal is 100 times higher owing to exhaled carbon dioxide in comparison with the usual concentration of ethylene in exhaled air. Similarly, at the 9R(30) line of CO_2 laser, the same concentration of CO_2 has an absorption coefficient equal to that of 1500 ppbV of NH_3. This value is also considerably higher (6 times) compared to the real range of breath concentration which is situated approximately at 250 ppb for ammonia.

Water vapor exhibits a broad continuum with occasional weak lines in the frequency range of the CO_2 laser (for H_2O at the 10P (14) laser line, $\alpha(H_2O)$ = 2.85×10^{-5} atm^{-1}cm^{-1}). The two dominant peaks are the absorption lines on 10R(20) and the most favorable one for ambient air measurement, the 10P(40) laser transition. A 5% concentration of H_2O has an absorption strength comparable to 46.9 ppbV of C_2H_4, that is the normal concentration of water in exhaled air has approximately the same influence in the photoacoustic signal as the normal concentration of ethylene.

Due to the additive character of the photoacoustic signal under normal pressure conditions, the presence of a large amount of water vapor and carbon dioxide impedes C_2H_4 detection in the low-concentration range (ppbV). Consequently, some means of selective spectral discrimination is required if ethylene is to be detected interference free in the matrix of absorbing gases. There are several ways to overcome this problem. One way is to remove CO_2 from the flowing sample by absorption on a KOH-based scrubber inserted between the sampling cell and the PA cell. Taking into account the nature of the specific chemical reactions involved in the CO_2 removal by KOH, a certain amount of water is also absorbed from the sample passing the scrubber. In this way, concentrations below 1 ppmV CO_2 (equivalent to a concentration of 0.07 ppbV of C_2H_4) can be achieved without influencing the C_2H_4 or NH_3 concentration.

Before entering the photoacoustic cell, the gas mixture passes through a KOH scrubber, which retains most of the interfering carbon dioxide. The removal of CO_2 is limited to the absorbent surface of the pellets. Hence, the larger the surface area or the more porous the granular solid, the larger the capacity of the system to absorb CO_2. At the same time, the flow resistance varies inversely proportional to the particle size. Large particles offer less resistance, but have the disadvantage of providing a smaller total area for reaction. The granules of KOH used were typically Merck KOH pellets GR for analysis, with approximate dimensions of 10x7x2 mm. When residence time (time of contact between CO_2 and absorbent) is less than 1 second, CO_2 absorption capacity is greatly reduced, so flow controllers were introduced in order to ensure this pre-requisite.

Potassium hydroxide is a caustic compound of strong alkaline chemical, dissolving readily in water, giving off much heat and forming a caustic solution. It is a white deliquescent solid in the form of pellets obtained by concentration of purified electrolytic potassium hydroxide solution with very low chloride content. It reacts violently with acid and it is corrosive in moist air toward metals such as zinc, aluminum, tin and lead, forming a combustible, explosive gas. It absorbs rapidly carbon dioxide and water from air. Cautions must be taken when used because the inhaled dust is caustic and irritant, and touching skin or clothes could lead to less or more severe chemical burnings.

The efficiency of the KOH scrubber was investigated using four recipients with different volumes (13 cm^3, 45 cm^3, 120 cm^3, and 213 cm^3, respectively), and was found out what type has to be used in order to reduce efficiently the amount of CO$_2$ from the exhaled air sample (Bratu et al., 2011). The KOH scrubber must neither change the ethylene concentration level, nor introduce new interfering gases. The measurements were made each time on the same person (healthy female, 30 years old) and with a new filling of KOH pellets. The gas from the sample bag was transferred into the PA cell at a controlled flow rate of 300 sccm (only for the 13 cm^3 trap) or 600 sccm, in order to ensure a sufficient time of flow in the scrubber column and to minimize any tendency for the vapor to stick to the cell walls or any other effects of internal outgassing of contaminants, which would otherwise lead to increase background signals during an experimental run. The typical resulting final pressure inside the PA cell was around 700 mbar and the corresponding responsivity was 170 cmV/W.

The experimental results without the KOH scrubber showed an equivalent ethylene absorption concentration of 2750 ppbV (with alveolar air collection) and 2350 ppbV (with mixt expiratory air collection), representing mainly the contribution of ethylene, carbon dioxide, water vapors and ammonia to the absorption of 10P(14) CO$_2$ laser line (Figure 2.8).

It was tested the efficiency of traps filled with KOH and having different volumes (between 13 cm^3 and 213 cm^3) in removing CO$_2$ from exhaled air. For the first measurement was used a trap with a small volume of 13 cm^3 of KOH scrubber, and was obtained a decrease of the PA signal down to 1-3 mV. The equivalent ethylene concentration was 435 ppb and 240 ppb, respectively (alveolar air collection vs. mixt expiratory air collection), indicating that the CO$_2$ concentration was reduced by factors of 6.3 and 9.8, respectively. Only in the case of this trap it was observed a peculiar behavior. Even if the laser power is constant, the PA signal and consequently the equivalent ethylene concentration increases in time after transferring the gas sample in PA cell. The increase of concentration starts from 50 ppbV and continues until it stabilizes at a level of 435/240 ppbV (after 10-15 minutes).

It is known that C_2H_4 (28.05 g/mol molar mass) is lighter than CO_2 (44.0099 g/mol molar mass). Because of that, after passing the KOH scrubber, first C_2H_4 enters in the PA cell and then CO_2 when the trap is no longer effective. So, at the beginning, was measured only the C_2H_4 concentration and then CO_2 starts to strongly interfere in absorption. It is possible that due to the geometry of the cell, a longer time is required in order to attain the total homogeneity of the molecules inside the resonant tube of the cell, but this is not advantageous for repeated measurements.

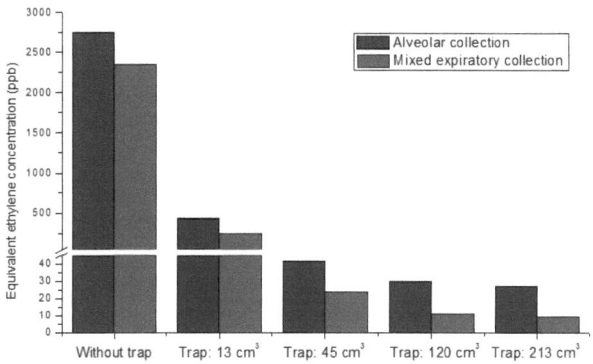

Figure 2.8. Efficiency of KOH traps for CO_2 removal from exhaled air.

Larger KOH traps proved to be more efficient in removal of CO_2 from the exhaled air. For the traps with volumes of 45 cm³, 120 cm³, and 213 cm³, respectively, the measured equivalent ethylene concentrations were 41.5/23.6 ppbV, 30/10.8 ppbV, and 26.8/9.1 ppbV, respectively. For larger traps (120 cm³ and 213 cm³), approximately same results were obtained, indicating that most of the CO_2 was removed. By using larger traps, a higher transfer rate of the gas mixture in the PA cell is possible, doubling the flow rate to 600 sccm.

For the two largest volumes, it was succeeded to reduce the CO_2 content from the exhaled air at a level influencing no more the C_2H_4 and NH_3 concentration values, fact proved by the constant evolution in time of all parameters. Therefore, the trap is effective only for a enough large amount of KOH pellets. It was found that a minimum volume of 120 cm^3 of KOH scrubber and a transfer rate of 600 sccm were optimum to insure the required efficiency.

Analyzing the four cases whit the scrubber, the dependence between the removed content of CO_2 and the used KOH quantity proved to be nonlinear, as one could expect. If is considered the content of the sample totally free of CO_2 after passing through the 213 cm^3 and 120 cm^3 KOH traps, is calculated a residual content of CO_2 in alveolar collection of 0.58% (5800 ppm) for the 13 cm^3 trap and of 0.016% (160 ppm) for the 45 cm^3 trap (less than half of the CO_2 concentration in the inhaled air).

The efficiency of the KOH scrubber is also calculated when it is used for multiple measurements (Figure 2.9). A clear saturation effect is evident: the KOH scrubber is not any more efficient when the same fill is used for multiple runs (it cannot absorb completely the CO_2 from the gas mixture). In the case of alveolar collection, the equivalent ethylene concentration increases by 2.3 times for the second run, by 2.6 times for the third run and by 3.4 times for the fourth run. For mixed expiratory collection, this saturation effect is even larger: the equivalent ethylene concentration increases by 2.4 times for the second run, by 8.5 times for the third run and by 20.2 times for the fourth run. The conclusion is that a new fill of KOH scrubber must be introduced after each measurement.

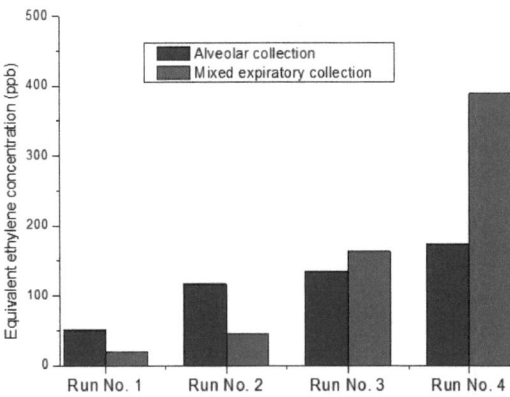

Figure 2.9. Decrease of KOH trap efficiency when the same fill was used for multiple measurements.

The lungs and airways are always moist, and inspired gas is rapidly saturated with water vapor in the upper segments of the respiratory system. The temperature in the airways and lungs is most identical with deep body temperature (approximately 37°C); at this temperature water vapor has a partial pressure of 47 torr (~6.2%). The increased saturation found at the third and fourth run for mixed expiratory collection is explained by a higher quantity of water vapors in exhaled breath (originating both from lungs and from upper segments of the respiratory system).

The nonlinearity of the CO_2 removal could be explained by the mechanism of the chemical reactions. First, the CO_2 combines with the water vapors present in the exhaled air in the form of carbonic acid:

$$CO_2 + H_2O \Rightarrow H_2CO_3. \qquad (7)$$

Further, the last one combines with the KOH, creating potassium carbonate and water, and releasing a small amount of heat:

$$H_2CO_3 + 2KOH \Rightarrow K_2CO_3 + 2H_2O + Energy. \qquad (8)$$

In the same time, K_2CO_3 is a highly hygroscopic compound with a retaining capacity of 0.2 g H_2O/1 g K_2CO_3, so the generated water will be only partially returned in the circuit.

The water is of high importance in limiting the rate of CO_2 absorption. High CO_2 concentrations entering the KOH absorber generates large quantities of water, because the reaction (8) is producing water. The absorption rate is greater thanks to the film of moisture coating the pellets and the same film impedes the access to the active potassium hydroxide pellet volume. More dedicated studies should be made in order to establish the moisture content for an optimum rate of absorption.

In conclusion, it was determined experimentally that in the process of CO_2 removal from the breath air samples, a quantity of minimum 120 cm³ KOH pellets should be used for a sampling bag of 750 mol in order to keep the detection of ethylene and ammonia traces free of CO_2 interference. It should be mentioned that this volume of 120 cm³ must be reconsidered for sample bags with a greater volume (> 750 mol) or when the gas transfer rate from the bag to the PA cell is larger (> 600 sccm).

Conclusions

This chapter describes an extremely sensitive apparatus based upon laser photoacoustic spectroscopy methods which can be used for the detection and measurement of trace gases at very low concentrations.

Two experimental set-ups were designed and characterized with the photoacoustic cell in an external configuration: the first one with a low power CO_2 laser where the saturation effects are negligible, and a second one with a high power CO_2 laser where the saturation effects are important and have to be taken into consideration. In the first case, the minimum detectable concentration was 0.9 ppbV (parts per billion by volume), while in the second case this parameter was improved to 0.29 ppbV. Comparing with the best results published previously in the literature, this minimum detectable concentration is better by a factor of 4.2 in the first case and by a factor of 13.1 in the second case. All measurements were done in ethylene with the 10P(14) line of a continuous wave CO_2 laser. This technology can dramatically impact detection in numerous areas.

Noises and other limiting factors which determine the ultimate detection sensitivity were presented, and various types of noises were measured.

The interfering gases play an important role both in limiting the sensitivity of the method and in the multicomponent analysis of the atmosphere. It was therefore discussed the experimental measures for reducing the influence of interfering gases in a single component measurement and the methods used in multicomponent analysis.

It was experimentally determined that in the process of CO_2 removal from the breath air samples, a quantity of minimum 120 cm³ KOH pellets should be used for a sampling bag of 750 mol in order to keep the detection of ethylene and ammonia traces free of CO_2 interference. It should be mentioned that this volume of 120 cm³ must be reconsidered for samples with a greater volume (> 750 mol) or in conditions of increasing the gas flow rate.

With the relevant characteristics of high sensitivity and specificity, laser photoacoustic spectroscopy holds a great potential for medical diagnostics.

References

Besson J. P. (2006). Photoacoustic spectroscopy for multi-gas sensing using near infrared lasers. These École polytechnique fédérale de Lausanne EPFL, n° 3670.

Bijnen F.G., Reuss J., Harren F.J.M. (1996). Geometrical optimization of a longitudinal resonant photoacoustic cell for sensitive and fast trace gas detection, *Rev. Sci. Instrum.,* Vol. 67, pp. 2914-2923

Bratu A.M., Popa C. , Matei, C. , Banita S., Dutu D.C.A. and Dumitras D.C. (2011). Removal of Interfering Gases in Breath Biomarker Measurements. *J. Optoelectron. Adv. Mater.,* Vol.13, No. 8, (August 2011), pp. 1045-1050, ISSN 1454-4164

Ca W., Duan Y., (2006), Breath Analysis: Potential for Clinical Diagnosis and Exposure Assessment, *Clinical Chemistry,* Vol 52, pp. 800-811

Cristescu S., Dumitras D.C. and Dutu D.C.A. (2000b). Ammonia and Ethene Absorption Measurements with a Tunable CO_2 Laser-Based Photoacoustic Trace Gas Detector. *Proc. SPIE ALT '99 International Conference on Advanced Laser Technologies,* Vol.4070, V.I. Pustovoy, V.I. Konov (Eds.), pp. 457-464, SPIE, ISBN 978-0-819-43707-5, Bellingham, WA, USA

Dumitras D. C., Bratu A. M. and Popa C. (2012). CO_2 Laser Photoacoustic Spectroscopy: Instrumentation and Applications, Intech, Croatia; *Chapter II in CO₂ Laser-Optimisation and Application,* ISBN 979-953-307-712-2, Ed. D. C. Dumitras b.

Dumitras D.C., Dutu D.C., Matei C., Magureanu A.M., Petrus M. and Popa, C. (2007). Laser Photoacoustic Spectroscopy: Principles, Instrumentation, and Characterization. *J. Optoelectr. Adv. Mater.,* Vol.9, No.12, (December 2007), pp. 3655-3701, ISSN 1454-4164

Dumitras D.C., Dutu D.C.A., Matei C., Cernat R., Banita S., Patachia M., Bratu A.M., Petrus M. and Popa C. (2011). Evaluation of Ammonia Absorption Coefficients by Photoacoustic Spectroscopy for Detection of Ammonia Levels in Human Breath. *Laser Phys.,* Vol.21, No.4, (April 2011), pp. 796-800, ISSN 1555-6611

Fink T., Büscher S., Gäbler R., Yu Q., Dax A., Urban W., (1996), An improved CO_2 laser intracavity photoacoustic spectrometer for trace gas analysis, *Rev. Sci. Instrum.* Vol. 67, pp. 4000-4004

Folke M., Cernerud L., Ekstrom M., Hok B., (2003), Critical review of non-invasive respiratory monitoring in medical care, *Med. Biol. Eng. Comput.* Vol. 41, pp. 377 - 383

Gerlach R. and Amer N.M. (1980). Brewster Window and Windowless Resonant Spectrophones for Intracavity Operation. *Appl. Phys.A,* Vol.23, No.3, (November 1980), pp. 319-326, ISSN 0947-8396

Harren F.J.M., Bijnen F.G.C., Reuss J., Voesenek L.A.C.J., Blom C.W.P.M., (1990a), Sensitive intracavity photoacoustic measurements with a CO_2 waveguide laser, *Appl. Phys. B,*Vol 50, pp. 137-144

Harren F.J.M., Reuss J., Woltering E.J., Bicanic D.D., (1990b), Photoacoustic measurement of agriculturally interesting gases and detection of C_2H_4 below the PPB level, *Appl. Spectrosc.* Vol. 44 pp. 1360-1368

Harren F.J.M., Reuss J., (1997), Spectroscopy, photoacoustic, in G. L. Trigg (Ed.), *Encyclopedia of Applied Physics* 19, VCH Publishers, New York, pp. 413-435

Nägele M., Sigrist M.W., (2000), Mobile laser spectrometer with novel resonant multipass photoacoustic cell for trace-gas sensing, *Appl. Phys. B,* Vol. 70 pp. 895-901

Pushkarsky M.B., Weber M.E, Baghdassarian O., Narasimhan L.R., Patel C.K.N., (2002), Laser-based photoacoustic ammonia sensors for industrial applications, *Appl. Phys. B* Vol. 75, pp. 391-396

Rooth R.A., Verhage A.J.L., Wouters L.W., (1990), Photoacoustic measurement of ammonia in the atmosphere: influence of water vapor and carbon dioxide, *Appl. Opt.* Vol. 29, pp. 3643-3653

Ryan J.S., Hubert M.H., Crane R.A., (1983), Water vapor absorption at isotopic CO_2 laser wavelengths, *Appl. Opt.*Vol. 22, pp. 711-717.

Siegman A.E. (1986). *Lasers*, University Science Books, ISBN 978-0-935-70211-3, Sausalito, CA, USA

Wang C., Sahay P., (2009), Breath Analysis Using Laser Spectroscopic Techniques: Breath Biomarkers, Spectral Fingerprints, and Detection Limits, *Sensors,* Vol. 9, pp. 8230-8262

Chapter 3.
Breath Ethylene Measurements

3.1. Introduction

Progressive medicine has been focused on the analysis of blood, urine, body fluid and tissue to improve disease identification or therapeutic approach and expressed a special interest in development of sensitive and adaptable trace gas analysis instruments. Because breath analysis are among the least invasive methods for monitoring a person's disease or exposure to a drug or an environmental pollutant, measurement of human biomarkers opens new fields of research (Navas et al., 2012). Exhaled breath air is a mixture of thousands of molecules, some of which are present at ppbV to pptV concentrations. These molecules, generated from endogenous (generated in the body) and exogenous (like inhaled air, ingested food and beverages, or any molecules that entered the body via extraneous routes) sources, provide a unique breath profile of pathophysiologic status (Risbyand Solga, 2006, Stepanov,2007). Potential advantages of breath test over other conventional medical tests include their non-invasive nature, low cost, and safety. This is an area where the modern day advances in technology and engineering meet the ever expanding need in medicine for more sensitive, specific and non-invasive test which makes this area a major front in the interface between medicine and engineering (Wang and Sahay, 2009).

The analysis of exhaled breath using LPAS is noninvasive and can be performed easily allowing a large number of patients to be studied. Most studies are focused on the detection of a single disease marker, as ethylene, present in human breath. The relation between a biomarker and a specific disease is often multi-fold. In some cases, a breath species is a biomarker that is indicative of more than one disease or metabolic-disorder; in other cases, one particular disease or metabolic disorder can be characterized by more than one chemical species. For instance, breath ethylene is an indicator of oxidative stress, lipid peroxidation (LP) inflammatory processes (chronic asthma, peritonitis), acute myocardial infarction, and it serves as a biomarker for radiation cell damage. Similarly, LP can be diagnosed through analysis of breath ethylene and pentane. These features require breath analysis to be not only highly-sensitive, but also highly-selective in order to obtain accurate information (Mashir and Dweik, 2009).

46

A significant biomarker present in exhaled breath is the hydrocarbon ethylene. Ethylene is known as a real time biomarker of oxidative stress status due to LP. Oxidative stress status is defined as the equilibrium between the formation and removal of free radicals. When the free radical capacity of the cell is overloaded, a complex chain of reaction occur leading to the destruction of cell membranes, and release of hydrocarbons. Several serious diseases in the body are manifestations of oxidative stress, where the loss of balance between free radical or reactive oxygen species (ROS) production and antioxidant systems is more pregnant, with strong negative effects on carbohydrates, lipids and proteins. The cell damage started by free radical action on biomolecules plays an important role in the pathogenesis of cancer, Alzheimer's, kidney or liver malfunction, asthma, neurological disorders and aging (Rahman,2007, Cao and Duan, 2006, McCurdy et al., 2007).

The oxidative modification (resulting in cell and tissue injury) of biological molecules is an essential part of normal biological activity. For example, radiation therapy uses ionizing radiation to kill cancer cells and shrink tumors. When considering ionizing radiations, a substantial part of the total interactions concerns water molecules, water being the major component of living tissue present in all biological systems. Consequently, mainly water ions and radicals are generated inside tissues as primary reactive species. Those reactive species (free radicals) then interact with biomolecules and damage them (indirect effect of radiation); in particular, they can start LP events on cell membranes. The ethylene produced as an effect of radiation-induced LP, in fact the process of free-radical-induced oxidative degradation of polyunsaturated fatty acids, diffuses through tissues inside the body, then it is collected by the haematic flow in the blood vessels, and transported to the lungs. The membranes separating the air in the lungs from the blood in the capillaries are very thin and are optimized for gas transport, so ethylene is easily emitted in exhaled breath. If the increase in the breath ethylene could be put in relation to the extent of LP events, then the monitoring of exhaled ethylene could be a simple method to follow up the increase of radiation damage in an organism (Giubileo et al., 2004, Wrixon et al., 2004).

Also, the cigarette smoke contains many toxic components that may induce ethylene formation. Ethylene oxide is a chemical product that induces cancer in the lungs. In order to monitor the damages caused by the inhaled smoke and inhaled vapor, a breath test is performed in order to obtain information on the volatile organic compounds under normal and stress circumstances (Rockville, 2010).

In this chapter, PA spectroscopy is being used to assess the amount of breath from smokers and patients under X-ray therapy via the exhaled ethylene biomarker.

3.2 Breath sample collection protocol

As far as the sampling procedure is concerned, it is used an extractive method, based on the collection of trace gas samples by some type of container or collecting medium and subsequent analysis in the laboratory. A problem may arise at this point due to some alterations of the gas composition caused by adsorption and desorption processes on the inner surface of the collecting container. The breath samples were obtained from volunteers who agreed to provide such samples at certain time intervals. The volunteers were asked to exhale into a sample bag with a normal exhalation flow rate. The breath samples were collected in 0.75-liter aluminum-coated bags (Quintron, Milwaukee, Wisconsin USA) equipped with valves that sealed them after filling (Figure 3.1).

Collecting a sample is a very easy procedure. After a normal inspiration, the volunteer places the Mouthpiece in his/her mouth, forming a tight seal around it with their lips. A normal expiration is then made through the mouth, in order to empty the lungs of as much air as required to provide the breath sample. For the mouth exhaled breath sample, the first portion of the expired air is directed into the discard bag (with the role in collection of the "dead-space" air: the first portion of an expired breath), while the alveolar air is diverted to the collection bag. When an adequate sample is collected, the subject stops exhaling and removes the mouthpiece/tube.

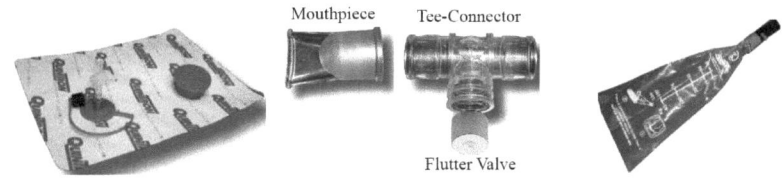

Figure 3.1
a) 0.75 L aluminum-coated bag;   b) Mouthpiece and tee-connector;   c) 0.40 L discard bag;

After the volunteer exhaled via the mouth the sample is collected, the gas from the sample is analyzed immediately or stored in the sample bag for later analysis. In either case, it is recommendable to seal the large port with the collection bag port cap furnished with the collection bag. The use of the port cap assures that the sample volume will not be lost due to a leak. Its use also avoids the contamination of the sample by gas diffusion through the one-way valve in the large port, if the sample is stored for a long period of time prior to its analysis (Popa et al., 2011).

The bags were inserted into a gas handling system (Figure 1.11), which ensured better control by means of two independently adjusted flow controllers of the upstream pressure and the flow rate through the sample bag. Breath samples are then directed to the sensitive PA cell, in which the gas is detected.

The experimental set-up is based on a line-tunable and frequency-stabilized CO$_2$ laser acting as a radiation source, because its emission spectrum overlaps with the absorption fingerprint of ethylene. It emits continuous wave radiation with an output power of 2-7 W and it is tunable between 9.2 and 10.8 μm on approximately 60 different vibrational-rotational lines. The laser beam is amplitude modulated by a mechanical chopper, then focused by a ZnSe lens and introduced in the PA cell. After passing the cell the laser beam power is measured by a digital powermeter. Its digital output is introduced in the data acquisition interface module together with the output from the lock-in amplifier. The gas concentration is proportional to the ratio between signal and laser power. All experimental data are processed and stored by a computer.

3.3 Ethylene in breath air following the inhalation of cigarette smoke

Cigarette smoke contains over 4700 chemicals and 69 of these are known to cause cancer. Smoking harms nearly every organ of the body and diminishes a person's overall health. Millions of people have health problems caused by smoking. Smoking compromises the immune system, making smokers more likely to have heart and blood vessels diseases, lung diseases and various types of cancer the most common being lung cancer. Smoking increases the risk of developing a number of other diseases. Many of these may not be fatal, but they can cause years of unpleasant symptoms.

About half of all smokers die from smoking-related diseases. If you are a long-term smoker, on average, your life expectancy is about 10 years less than a non-smoker. Many smoking-related deaths are not quick deaths. For example, if you develop COPD you can expect several years of illness and distressing symptoms before you die.

The risk of premature death and the chance of developing cancer from smoking cigarettes depend on many factors, including the number of years a person smokes, the number of cigarettes he or she smokes per day, the age at which he or she began smoking, and whether or not he or she was already ill at the time of quitting. For people who have already developed cancer, quitting smoking reduces the risk of developing a second cancer (Nee, 2013).

Cigarette smoke contains several reactive oxygen species (ROS) which may damage lipids, proteins, DNA, carbohydrates and other biomolecules. Most ROS have a short half-life and cause damage locally except H_2O_2 which has a relatively long half-life and can travel long distances causing DNA damage at distant sites. Increased production of ROS leads to an imbalance between the oxidative forces and the antioxidant defense systems, favoring an oxidative stress or injury. ROS can influence cell survival and genomic stability (Hecht, 1999).

Cigarette smoke contains various substances which stimulate the increased productivity of free radicals *in vivo* and thus disturb the oxidant: antioxidant homeostasis. Free radicals after binding to macromolecules oxidize them and lead to the production of adducts which interfere with normal cellular processes.

The cigarette smoke contains many toxic components (heavy metals, free radicals, chemicals) that may induce LP in the lung epithelium (Dumitras et al., 2008; Giubileo et al., 2004). LP is the free-radical-induced oxidative degradation of polyunsaturated fatty acids. Biomembranes and cells are thereby disrupted, causing cell damage and cell death. LP generates alkenes such as ethane, pentane, which are eliminated in the breath. As a marker of free-radical-mediated damage in the human body, the measurement of the exhaled volatile hydrocarbons, such as C_2H_4, is a good noninvasive method to monitor LP.

Ethylene is dangerous for smokers because ethylene oxide is a chemical product that induces cancer in the lungs. For the moment, it is difficult to separate the exogenous and endogenous origin of the ethylene in the smoker's breath. The intensity of smoke induced oxidative damage varies with the degree or frequency of exposure to cigarette smoke.

To monitor the damages caused by the inhaled smoke, a breath test was performed using LPAS, which gives information on volatile compounds, such as ethylene. The method has been applied in studying how the concentration of the ethylene coming out from human lungs is modified after smoking. The evolution of ethylene concentration was followed up in the case of nonsmokers and smokers.

The breath samples analyzed were obtained from volunteers who agreed to provide such samples at certain time intervals. Measurements were made at 10P(14) CO_2 laser line (10.53 μm), where the ethylene absorption coefficient has the maximum value (30.4 cm^{-1}atm^{-1}).

For healthy nonsmoking young subjects the average of exhaled concentration is 0.017 ppmV. In case of smoker persons the difference was obvious (Figure 3.2). The ethylene concentration increases till 1.6 ppmV, presenting values at least 100 times higher for smokers than those registered for the nonsmokers.

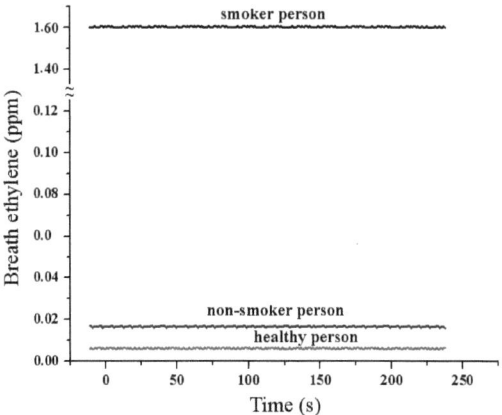

Figure 3.2. Ethylene content of exhaled air from nonsmokers and smokers persons

It's clear that cigarette smoke contains a large amount of free radicals and lungs can be affected (lung cancer). The performed breath test gives us important information about the toxicological implications of free radicals caused by the inhaled smoke.

3.4 Ethylene in breath air following x-ray therapy

Radiation therapy (also called radiotherapy, X-ray therapy, or irradiation) is the use of certain type of energy (called ionizing radiation) to kill cancer cells and shrink tumors. Radiation therapy injures or destroys cells in the area being treated by damaging their genetic material, making it impossible for these cells to continue grow and divide. Although radiation damages both cancer cells and normal cells, most normal cells can recover from the effects of radiation and function properly. The goal of radiotherapy is to damage as many cancer cells as possible, while limiting harm to nearby healthy tissue (Naidu, 2010).

Radiation therapy is commonly applied to the cancerous tumor because of its ability to control cell growth. Ionizing radiation works by damaging the DNA of cancerous tissue leading to cellular death (Reynolds and Schecker, 1995). To spare normal tissues (such as skin or organs which radiation must pass through to treat the tumor), shaped radiation beams are aimed from several angles of exposure to intersect at the tumor, providing a much larger absorbed dose there than in the surrounding, healthy tissue. Besides the tumor itself, the radiation fields may also include the draining lymph nodes if they are clinically or radiologically involved with tumor, or if there is thought to be a risk of subclinical malignant spread. It is necessary to include a margin of normal tissue around the tumor to allow for uncertainties in daily set-up and internal tumor motion. These uncertainties can be caused by internal movement (for example, respiration and bladder filling) and movement of external skin marks relative to the tumor position.

Radiation doses for cancer treatment are measured in a unit called a gray (Gy), which is a measure of the amount of radiation energy absorbed by 1 kilogram of human tissue. Different doses of radiation are needed to kill different types of cancer cells (Nee, 2013).

Radiation therapy can cause both early (acute) and late (chronic) side effects. Acute side effects occur during treatment, and chronic side effects occur months or even years after treatment ends. The side effects that develop depend on the area of the body being treated, the dose given per day, the total dose given, the patient's general medical condition, and other treatments given at the same time.

The effect of ionizing radiation on living cells is supposed to modify the oxidative stress status in the human body through an increase in the peroxidation processes started by the free water radicals generated by indirect radiation effect in living tissue. Important events of the peroxidation take place in the cell membranes determining the release of small linear hydrocarbon molecules through the lipid peroxidation pathways. A fraction of the hydrocarbon molecules generated in the tissue (one among them is the ethylene) will be transported to the lungs by the blood and release in the exhaled breath.

Studying the effects of ionizing radiation on biological materials, experiments were made to evaluate the breath emission of ethylene as a result of LP in human body. Exhaled air eas analyzed from healthy persons and from patients with cancer receiving radiation treatment based on X-ray external beam after tumor surgery.

Exhaled air was analyzed from 6 patients between 32 and 77 years old receiving radiation treatment based on X-ray external beam after malign tumor surgery

(Dumitras et al., 2008). The patients received fractional doses as high as 2 to 8 Gy depending on the cancer type. For this experiment patients were asked to exhale into sample bags at a normal exhalation flow rate. Breath samples were taken at certain time intervals (before and after X-ray therapy) were transferred in the PA cell and were analyzed in the continuous nitrogen flux. The KOH trap inserted in the gas circuit is used to remove as much as possible the high quantity of CO_2 from the exhaled air. Each sample was analyzed by LPAS and the PA signal was elaborated to evaluate the ethylene concentration contained inside the sample.

Measurements were made at 10P(14) CO_2 laser line where the laser output power is 5 W.

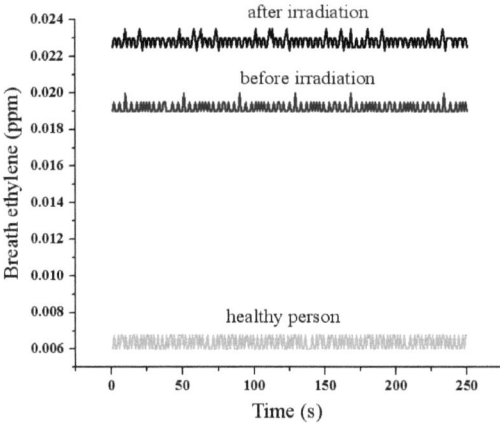

Figure 3.3. Ethylene content of exhaled air from patients receiving X-ray treatment

For a healthy person exhaled ethylene was found at a level of 0.006 ppmV, whereas for patients receiving X-ray therapy before irradiation ethylene levels were at 0.019 ppmV and after irradiation at 0.023 ppmV (Figure 3.3).

As a first observation, after the X-ray irradiation the ethylene concentration rises, showing that LP took place. So, it is possible to detect the process in the very first minute after irradiation. The effect of LP is more powerful on the cancer cells, while the healthy cells even affected have higher recovery ability. A surprising decrease in the level of ethylene concentration was observed in the exhaled air after 15 minutes, the level being even lower than the normal level of the patient (e.g. the level measured before any irradiation). This could be explained as a body reaction to the increased level of peroxidic attack: higher the rate of damage, higher the self-defense response of the human organism. Further work is required in order to verify this hypothesis.

However, after X-ray therapy the ethylene concentration rises, showing that LP took place, being possible to detect the process in the very first minute after irradiation. The effect of LP is more intense on the cancer cells, while the healthy cells even affected have higher recovery ability.

Professor Nic Jones, Cancer Research UK's chief scientist, said: 'Around half of all cancer patients are given radiotherapy and it has been at the heart of helping improve survival rates so that today one in two cancer patients will survive for at least ten years. Doctors and researchers are constantly looking for ways to improve treatments and this approach could open the door to a whole new way of giving radiotherapy.

Conclusions

In this chapter, both the feasibility and the importance of monitoring exhaled ethylene from different patients have been shown. This gas, a biomarker of lipid peroxidation processes, has been measured using a CO_2-laser-based photoacoustic spectrometer.

The large amount of free radicals contained in cigarette smoke is probably the cause for the lung damage. The test for a smoker person shows a higher level of ethylene compared to nonsmokers and the level of the CO_2 in the expired breath alters considerably the true C_2H_4 concentration and the trap starts to be effective for a large amount of KOH pellets.

Ethylene concentration was also determined in the air exhaled from patients affected by cancer and treated by external radiotherapy. The breath test based on the laser photoacoustic analysis of ethylene might help in monitoring the radiotherapy treatment toxicity.

Measuring human biomarkers in exhaled breath is expected to revolutionize diagnosis and management of many diseases and may soon lead to rapid, improved, lower-cost diagnosis, which will in turn ensure expanded life spans and an improved quality of life.

References

Biggins J., Renfree S., (2002), The hazards of surgical smoke are not to be sniffed at! *Journal of Peri-Operative Nursing*

Cao W., Duan Y., (2006) Breath analysis: potential for clinical diagnosis and exposure assessment, *Clin. Chem.* Vol. 52, pp 800-811

Dumitras D.C., Dutu D.C., Matei C., Magureanu A.M., Petrus M., Popa C. and Patachia M. (2008). Measurements of Ethylene Concentration by Laser Photoacoustic Techniques with Applications at Breath Analysis. *Rom. Rep. Phys.,* Vol.60, No.3, pp. 593-602, ISSN 1221-1451

Ferenczy A., Bergeron C. and Richard R.M. (1990). Human papillomavirus DNA in CO₂ laser-generated plume of smoke and consequences to the surgeon, *Obstet Gynecol*, Vol74, No.6, pp. 950-954

Giubileo G., Puiu A., Argirò G., Rocchini P., Borra E., (2004). Analysis of the breath from patients treated by anti-tumour radio-therapy", *Laser Physics,* Vol. 14, No. 2, pp. 243–249

Giubileo G., Puiu A. and Dumitras D.C., (2004). Detection of Ethylene in Smokers Breath by Laser Photoacoustic Spectroscopy. *Proc. SPIE ALT'03 International Conference on Advanced Laser Technologies: Biomedical Optics,* Vol.5486, R.K. Wang, J.C. Hebden, A.V. Priezzhev, V.V. Tuchin (Eds.), pp. 280-286, SPIE, ISBN 978-0-819-45418-8, Bellingham, WA, USA

Hecht S.S. (1999). Tobacco Smoke Carcinogens and Lung Cancer, *JNCI J Natl Cancer Inst* Vol. 91 No.14, pp. 1194-1210

Mashir A., Dweik R.A. (2009). Exhaled breath analysis: The new interface between medicine and engineering, *Adv. Powder Technol.* Vol.20, pp. 420-425

McCurdy M.R., Bakhirkin Y., Wysocki G., Lewicki   (2007) Recent advances of laser-spectroscopy-based techniques for applications in breath analysis, *J. Breath Res.* Vol.1

Naidu K. M. (2010). Community Health Nursing, Published by Gennext Publication, ISBN:978-81-908675-1-1

Navas M.J., Jimenez A.M., Asuero A.G., (2012). Human biomarkers in breath by photoacoustic spectroscopy, *Clinica Chimica Acta.* Vol. 413, No. 15, pp. 1171-1178

Nee P.W. (2013). The Key Facts on Cancer: Everything You Need to Know About Cancer, Published by MedicalCenter.com, ISBN-10: 1484825128

Ott D.E., (1993) Smoke production and smoke reduction in endoscopic surgery: preliminary report, *Endosc Surg Allied Tech.*

Popa C., Bratu A.M., Matei C., Cernat R., Popescu A.and Dumitras D.C. (2011b). Qualitative and Quantitative Determination of Human Biomarkers by Laser Photoacoustic Spectroscopy Methods. *Laser Phys.,* Vol.21, No.7, pp. 1336-1342, ISSN 1555-6611

Rahman K. (2007) Studies on free radicals, antioxidants, and co-factors, *Clinical Interventions in Aging,* Vol.2, pp. 219–236

Reactor Concepts Manual, USNRC Technical Training Center, Biological Effects of Radiation, pp. 1-23

Reynolds R.J. and Schecker J.A. (1995). Radiation, Cell Cycle and Cancer, Book Id: WPLBN0000220552

Risby T.H., Solga S.F. (2006). Current status of clinical breath analysis, *Appl. Phys. B,* Vol. 85, pp. 421-426

Rockville M.D. (2010). How tobacco smoke causes disease: the biology and behavioral basis for smoking-attributable disease, *U.S. Department Of Health And Human Services* Public, Health Service Office of the Surgeon General, pp.1-9

Smith J., Yeh Hsu-Chi, Muggenburg B., Guilmetter R., Martin L.S., and Strine P.W., (1992). Study design for the characterization of aerosols during surgical procedures, *Scandinavian Journal of Work Environmental Health 2-* 2

Stepanov E.V.(2007). Methods of highly sensitive gas analysis of molecular biomarkers in study of exhaled air, *Physics of Wave Phenomena,* Vol. 15, pp. 149-181

Wang C., Sahay P. (2009). Breath analysis using laser spectroscopic techniques: breath biomarkers, spectral fingerprints, and detection limits, *Sensors,* Vol. 9, pp. 8230-8262.

Winston C. (1994). The effects of smoke plume generated during laser and electrosurgical procedures, *Minimally Invasive Surgical Nursing.*

Wrixon A.D., Barraclough I., Clark M.J., Ford J (Ed), Diesner-Kuepfer A., Blann B. (2004) Radiation, people and the environment, *IAEA,* pp. 1-85

Chapter 4.
Quantitative analysis of laser surgical smoke:targeted study on eight toxic compounds

4.1 Introduction

The smoke generated during surgical procedures has been studied for many years in an effort to understand what hazards and risks pose to people. In particular, operating room personnel (OR) are exposed to this surgical smoke every day, spending significant hours in a toxic environment. Assessing the qualitative and quantitative composition of surgical smoke will give the possibility to better understand the risks associated with long term exposure.

Surgical smoke and aerosols are known to be byproducts of the interaction between biological tissues and laser beams, electro surgical units, drills, ultrasonic scalpels. These energy based instruments are dedicated to cut, coagulate, ablate, and vaporize biological tissue by generating heat. Lasers and electro surgical units have the same mechanism, rising the cells temperature above the boiling point, causing membrane rupture and dispersing similar types of particles, thus creating the same type of surgical smoke (Ulmer, 2008). In the case of ultrasonic scalpels, the tissue is vaporized at lower temperatures, without a burning process, and therefore aerosols are mainly created (Ott et al., 1998).

Previous qualitative studies of surgical smoke have shown that the composition considerably depends on the type of surgical device, nature of tissue, and duration of surgery(Al Sahaf et al., 2007, González-Bayón et al., 2006, Travella, 2001). Water vapor is the main component estimated to represent 95%, although exact proportion obviously depends on the nature of tissue. This water vapor acts as a transport mean for the other 5% cellular debris, such as chemicals, blood and tissue fragments, also viruses and bacteria that present a potential hazard [Al Sahaf et al., 2007, González-Bayón et al., 2006].

The technique used to create surgical smoke is very important, being directly related to the size of particles. The smallest particles are created by electrocautery, with the mean size of 0.07 µm. Laser tissue vaporization creates larger particles, with the mean size of 0.31 µm, and ultrasonic devices creates particle between 0.35 and 6.5 µm (Ulmer, 2008) The smaller the particles are, the further distances may travel into the body and produce more cellular damage. The importance of particle size was demonstrated in a study showing that particles smaller than 100 µm remain airborne, particle of 5 µm or larger may deposit in the upper airway and particles smaller than 2 µm in size can easily be deposited in the alveoli, producing lung damages (Travella, 2001).

The chemical byproducts of surgical smoke have been identified in a list presented in a review by *Barrett and Garber* (Barrett and Garber, 2003). The most prominent chemicals found in electrocautery smoke were hydrocarbons, phenols, nitriles, and fatty acids (Hensmanet al., 1998). Many of these chemicals are of significant concern, and others, including acrolein (Hensmanet al., 1998) and benzene (link:ofmpub.epa.gov) have already been identified as carcinogens. Repeated exposure to these chemicals is associated with considerable health hazard. *Alp et al.* shown that surgical smoke may be responsible of signs of acute or chronic inflammation of the respiratory tract, headache, asthenia, nausea, cardiovascular dysfunction, leukemia, hepatitis, HIV infection, irritation of to the eyes and throat (Alp et al., 2006).

The exposure to surgical smoke can aggravate existing lung diseases or can develop new pulmonary or cardiovascular disorders. Transmission of infection by viruses is also important risk (Garden et al., 1988, Garden et al., 2002). These potential dangers to human health depend on many factors such as exposure mode, intensity and duration of exposure, even on each individual gene.

The quantitative data of surgical smoke composition have been partially studied. Reproducible results are difficult to obtain because of the multiple products present in diverse forms in surgical smoke. However, latest scientific publications in the field have focused on measuring chemicals concentrations in order to establish the risks for OR personnel when they are exposed to surgical smoke (Moot et al., 2007, Gianella and Sigrist., 2010, Gianella, and Sigrist, 2012 , Petrus et al., 2012). Gas components from the surgical smoke were detected with different methods like ion flow tube-mass spectroscopy (SIFT-MS) (Moot et al., 2007), gas chromatography-mass spectroscopy (GC-MS) (Weston et al., 2009, Chung et al., 2010) and mid-infrared difference-frequency-generation laser spectroscopy (MIR-DFG) (Gianella and Sigrist, 2012), but despite a very good selectivity and sensitivity, the calibrations are elaborated processes (for the first two methods) or the detection limits remain in the ppm range. One method that combines high sensitivity and selectivity with an easy calibration, and other advantages like multicomponent capability, real time data analysis or no need for sample preparation is the LPAS. This method was used for monitoring various surgical smoke constituents in order to understand the toxic impact on human health.

In this chapter, a study of the quantitative composition of surgical smoke obtained *in vitro* by irradiating fresh animal tissues with a CO_2 laser, in nitrogen atmosphere, is presented. The study employs LPAS to determine traces of acetonitrile, acrolein, ammonia, benzene, ethylene, toluene, carbon dioxide and water vapors from surgical smoke. The laser power, exposure time, and type of tissue are the three parameters that influence the gas concentrations. The measurements results were compared with the exposure limits given by health organizations for the purpose of assessing the health risk linked to smoke exposure.

As the exposure to surgical smoke can last from minutes to several hours (e.g. tumor resection), an extensive knowledge about the smoke composition helps in evaluating long-term exposure hazards and the summing effects of multiple noxious compounds.

4.2 Protocol for smoke sample measurements

The smoke samples were produced *in vitro* by laser tissue vaporization in nitrogen atmosphere and were investigated by a CO_2 laser photoacoustic spectroscopy system. The photoacoustic system is illustrated in Figure 4.1

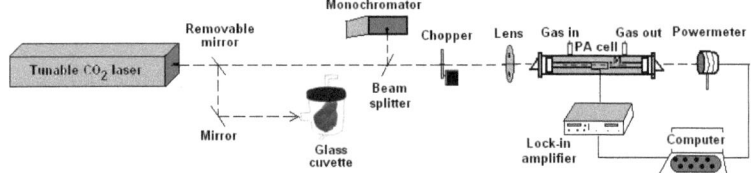

Figure 4.1. Schematic description of the photoacoustic detection chain

As a radiation source, it was used a commercial, cw, line-tunable CO_2 laser with output power till 50 W and tunable on 73 different lines between 9.2 – 10.8 μm (Coherent GEM SELECT 50™ laser). The minimum adjustable output power is about 10 W, since below this value the laser beam presents fluctuations. This power level is satisfactory for the tissue vaporization, but it produces saturation effects inside the PA cell, leading to erroneous measurements. To avoid the saturation effects, the laser power was minimized by introducing a beam splitter in the path of the laser beam. The ZnSe beam splitter is positioned at a distance of approx. 30 cm of the laser output and is placed at an angle of 45^0 to the direction of propagation of the laser beam. It has a 25 mm diameter, with a 60% transmission and 40% reflection. The reflected beam was coupled into a monochromator MacKen CO_2 Spectrum Analyzer (Model 16-A) for a precise laser line setting. A real-time monitoring powermeter Vega Ophir (with F150A-V1 head) was used to adjust the optimum laser power. Behind the beam splitter, the light beam was modulated by a mechanical chopper (DigiRad C-980 or C-995) operating at the appropriate resonant frequency of the cell (564 Hz), focused by a ZnSe lens (f = 400 mm), and introduced in the PA cell. An optimum acoustically resonant PA cell is used with a responsivity of 433 cmV/W (Bratu et al, 2013). Inside the cell, the pressure waves, generated when the laser radiation is absorbed by the trace gas molecules, are measured with sensitive miniature microphones (Knowles electret - EK-23024). After the PA cell, the power of the laser beam is measured by a laser powermeter (Rk-5700 from Laser Probe Inc). Its digital output is introduced in the data acquisition interface module together with the output from a lock-in amplifier (Stanford Research Systems model SR 830), which filters and amplifies the signal from microphones. The gas concentration is proportional to the ratio between signal and laser power. All experimental data are processed and stored by a computer. It has to be mentioned that this experimental set-up is one of the most sensitive system reported till now in the literature, allowing measurement of trace gas concentrations at a level of tens of ppt (Bratu et al., 2013).

The same CO_2 laser was used for the irradiation of the animal tissue samples, considering the 10P(14) laser line as reference. Smoke samples were produced in a glass cuvette, which allows vaporization of small volumes of tissue with the CO_2 laser in a selected atmosphere. It is fitted with a rubber stopper in order to offer isolation against outside contamination. A glass tube with a ZnSe window was attached allowing the laser beam to enter the cuvette and produce surgical smoke from different samples. Two gas connections permit to flush the buffering gas (nitrogen or synthetic air), which transports the smoke samples to the PA cell. After irradiation the removable mirror clears the optical path to permit spectroscopic measurements.

The resulted smoke is then transferred by a nitrogen flow into the PA cell, where the desired gases were detected. Due to the size of PA cell, the measurements correspond to a dilution of the smoke sample in 1 dm³ of nitrogen.

The measurements enabled the investigation of eight gases of major interest such as acetonitrile, acrolein, ammonia, benzene, ethylene, toluene, carbon dioxide, and water vapor. These gases possess strong characteristic absorption features in the wavelength range of the CO_2 laser, so eight laser lines were chosen which correspond to the highest absorption coefficient for each gas (see Table 4).

Table 4. Gases detected in the CO_2 laser wavelength range

Gas	CO_2 laser line	Wavelength [µm]	Absorption coefficient [cm⁻¹ atm⁻¹]
Acetonitrile	9P(16)	9.517	0.15
Acrolein	10R(14)	10.286	2.80
Ammonia	9R(30)	9.217	56.00
Benzene	9P(30)	9.636	2.00
Ethylene	10P(14)	10.529	30.40
Toluene	9P(28)	9.618	0.67
Carbon dioxide	9R(18)	9.3	3×10^{-3}
Water vapor	10R(20)	10.24	8.3×10^{-4}

The transfer of the smoke sample to the PA cell is performed using a vacuum/gas handling system that ensures the PA cell purity. The vacuum system provides evacuation of the entire gas handling system, including the PA cell, either totally or in different sections. The gas or gas mixture is introduced at controlled flow rates within a broad range (10-1000 sccm). The process permits rinsing the PA cell or the gas handling system with pure nitrogen, calibrating the PA spectrometer with a certified gas mixture, or the admission of the sample and carrier gas on different paths. A number of pressure gauges allow gas pressure reading inside the PA cell, as well as in different segments of the system (Figure 4.2).

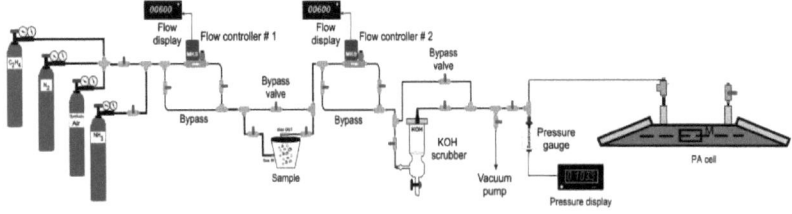

Fig 4.2. Flow system for handling gas samples

The sample cuvette is safely inserted in the detection chain, allowing the tissue irradiation, the transfer of the smoke inside the PA cell, and its subsequent removal from the optical path. The non-absorbing gases acting as carrier is nitrogen. The trace gas concentration in the sample gas is quickly analyzed ensuring a response time on the order of minutes or even seconds.

4.3 Quantitative study of six toxic compounds from surgical smoke

For the *in vitro* investigations four different types of pig tissues samples were used: muscle, kidney, heart, and skin. These samples were fresh purchased from the local slaughterhouse and they were cut in slices using the external surface of material as the preferential irradiation target to ensure as homogenous site as possible. The fresh animal tissues were positioned in the special cuvette which allows smoke production in a controlled nitrogen atmosphere. The tissues were irradiated at 10 and 15 W laser powers and for time lengths of 5 s, 10 s and 15 s.

The smoke samples from the cuvette were transferred by a nitrogen flow into the PA cell at a controlled flow rate of 600 sccm, in order to ensure sufficient time of flow in the scrubber column to remove interfering gases. The absorbing gases which strongly interfere in the detection method are water vapors and carbon dioxide which are filtered using a KOH scrubber. A quantity of minimum 120 cm³ KOH pellets was used to keep the detection traces free of interference, while the fill of KOH scrubber was renewed after each measurement. Measurements were realized at the atmospheric pressure (1024-1034 mbar) and room temperature.

After tissue vaporization, acetonitrile (C_2H_3N), acrolein (C_3H_4O), ammonia (NH_3), benzene (C_6H_6), ethylene (C_2H_4), and toluene (C_7H_8) were detected. It was observed that their concentration strongly depends on the type of tissue, laser power and exposure time (Table 5).

Table 5. Quantitative results of chemical compounds in surgical smoke obtained from CO_2 laser vaporization of fresh animal tissues

Type of pig tissue	Power [W]	Exposure time [sec]	C_2H_3N [ppm]	C_3H_4O [ppm]	NH_3 [ppm]	C_6H_6 [ppm]	C_2H_4 [ppm]	C_7H_8 [ppm]
Skin	10	5	351.5	32	23.15	31.75	0.395	73.25
Skin		10	140	42.2	37.2	13.8	0.578	32.8

Skin		15	235	59.4	39	16.6	0.632	31.7
Skin		5	399	36	29.4	38.6	0.403	73.7
Skin	15	10	429	47.1	35.2	44.4	0.486	134
Skin		15	285	76	51.6	26.4	1.19	77
Muscle		5	130	25.4	21.5	9.27	0.241	28.8
Muscle	10	10	153	33.4	23.7	10	0.320	30.8
Muscle		15	179	43.2	24.4	11.43	0.336	35.1
Muscle		5	163.2	29.4	22.9	11.02	0.263	30.2
Muscle	15	10	170	42.8	30.5	13.59	0.506	31.8
Muscle		15	183	44.6	30.8	15.48	0.514	37
Heart		5	117.2	25	19.8	19.8	0.201	70.03
Heart	10	10	188	27.13	21.5	21.5	0.313	44.97
Heart		15	293	24.5	18.93	18.93	0.368	42.37
Heart		5	205	15.93	14.23	14.23	0.180	48.47
Heart	15	10	166	15.90	12.73	14.7	0.217	35.3
Heart		15	148	14.2	12.3	14.63	0.283	33.93
Kidney		5	185	16.25	14.15	63.25	0.179	55.15
Kidney	10	10	109	23.55	17.95	9.71	0.264	23.55
Kidney		15	108.85	39.05	26.65	9.37	0.485	23.65
Kidney		5	61.65	16.4	17.65	5.43	0.258	12.85
Kidney	15	10	125.45	27.1	18.9	8.58	0.344	18.7
Kidney		15	223	55.8	31.9	29.35	0.839	50.8

The results listed in Table 5 were obtained when tissue samples were irradiated under the same conditions. Smoke is strongly affected by laser parameters that influenced the emitted substances in a characteristic manner. This gives the possibility to recognize special conditions of interaction by monitoring selected substances. The laser–tissues interactions are dominated by photochemical processes, but on a different temperature level for different interaction times.

Gas concentrations depend on the amount of vaporized tissue which is related to laser power and exposure time. The laser power was set at 10 and 15 W, while the exposure time was selected at 5, 10 and 15 s. The influence of laser power and exposure time for the muscle samples can be seen on Figure 4.3.

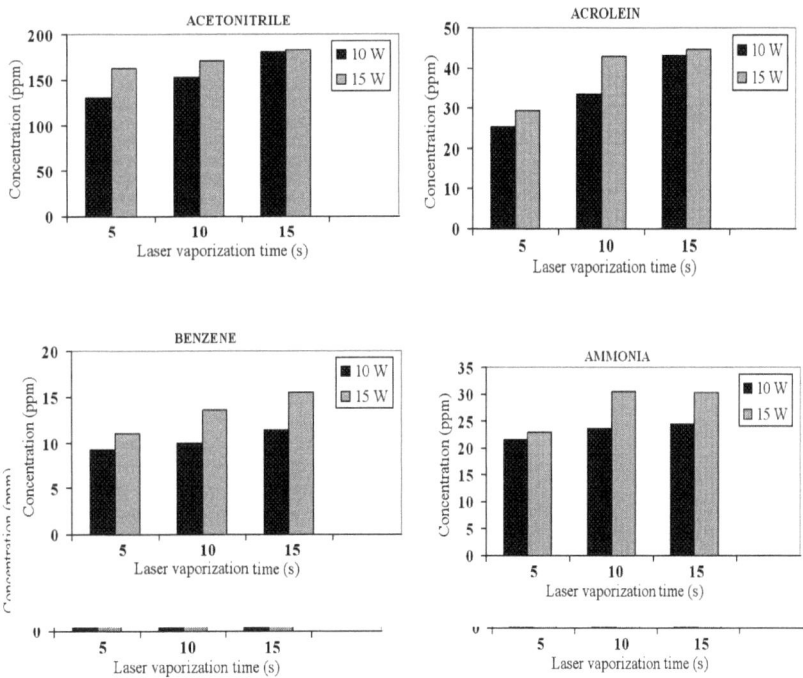

Figure 4.3. Mean values of gas concentration for the muscle samples

The results show that the concentration for all gases increases with laser power and exposure time, ranging from sub-ppm level for ethylene to almost 200 ppm for acetonitrile. Also, it can be observed that the increase in concentration is not linear with the laser power, nor the exposure time. The presented concentrations are the mean values averaged over all measurements.

The influence of the tissue type on the gas concentration is represented in Figure 4.4. Smoke samples were obtained by the irradiation of tissues, maintaining the same parameters: laser power of 10, 15 W and an exposure time of 15 s.

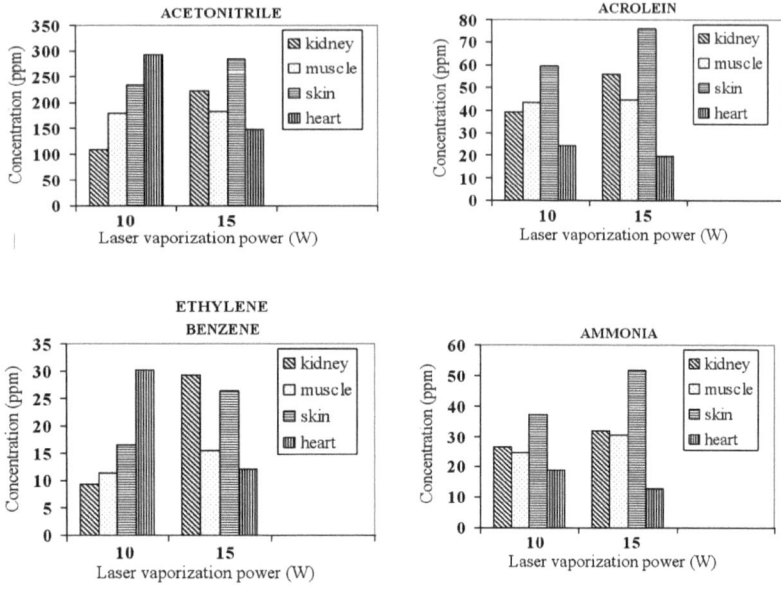

Figure 4. 4. Mean values of gas concentration for the skin, muscle, heart, and kidney samples

The results showed that for kidney, muscle and skin samples, the concentrations for all gases increase with laser power. Unexpectedly, the heart samples present a different behavior after the CO_2 laser vaporization. In this case, for all six gases the concentration decreases with the laser power. One explanation can be given by the complex nature of the heart tissue, generally very rich in blood vessels. It can be assumed that after vaporization, more blood particles are released and less chemicals.

International health organizations have developed guidelines and recommendations to protect healthcare workers from the negative effects of surgical smoke. These guidelines and recommendations include measured chemicals commonly found in surgical smoke. The values given by the United States of Occupational Safety and Health Administration (OSHA) set the permissible exposure limits (PEL) for workers exposure to specific chemicals usually given as time-weighted average over 8 hour period (TWA) and short-term exposure limits for 15 minutes period (STEL) (www.osha.gov). OSHA does not have specific regulation related to the evacuation of surgical smoke. The National Institute for Occupational Safety and Health (NIOSH) investigates potential occupational health risks and makes recommendations to OSHA (www.cdc.gov). The recommendations of NIOSH are referenced on the OSHA website on smoke evacuations.

The average concentrations determined by the measurements were: acetonitrile - 190 ppm, acrolein - 35 ppm, ammonia - 25 ppm, benzene 20 - ppm, ethylene - 0.41 ppm, and toluene - 45 ppm but they vary from sample to sample. Comparing these measurements with the REL and PEL for TWA and STEL (Table 6), it was observed that acetonitrile, acrolein, and benzene were found in concentrations exceeding these recommendations (see Table 6).

Table 6. Comparison of the known (NIOSH, OSHA) established exposure limits and experimental results

Gas	Permissible exposure limit (ppm)		Recommended exposure limit (ppm)		Average concentrations measured (ppm)
	TWA	STEL	TWA	STEL	
Acetonitrile	40	60	20	20	190 !
Acrolein	0.1	0.3	0.1	0.3	35!
Ammonia	35	35	25	35	25
Benzene	1	5	0.1	1	20!
Ethylene	-	-	200	200	0.41
Toluene	200	300	100	200	45

Although acetonitrile has a modest toxicity in small doses when exceeding the REL by 9.5 times, it can be considered to have toxic effects for humans. Acetonitrile is metabolized in the body to produce hydrogen cyanide and thiocyanate, which are responsible for acetonitrile hazard. The U.S. Environmental Protection Agency (EPA) noted that acetonitrile and its metabolites are absorbed by the blood. Studies on animals revealed that following acetonitrile inhalation, parent compound or metabolites were found in the brain, heart, liver, kidney, spleen, blood, stomach, and muscle. In the case of humans, following acute inhalation, metabolites were also found in the above mentioned organs as well as skin, lungs, intestine, and urine [www.epa.go]. In 2012, a study on acetonitrile properties, exposure, metabolism and toxicity noted that acute symptoms are usually abdominal pain, convulsions, labored breathing, weakness, unconsciousness and redness in the skin and eyes (Cleverson et al., 2012). With prolonged exposure, the liver, lungs, kidneys, and central nervous system may be affected.

Significant levels of acrolein were found even 350 times higher related to REL for TWA. This concentration is problematic from the healthy perspective of the exposed personnel, taking into account that EPA classifies acrolein as a group C carcinogen for adrenal gland cancer in rats (www.epa.gov/iris/subst/0364.htm). Acrolein presents high reactivity and is rapidly metabolized by human organism following inhalation, oral or dermal exposures. It acts as lipid peroxidation byproduct leading to various human diseases (Feng et al., 2006). High levels of acrolein may also affect respiratory, reproductive, neurological and hematological systems. In a study, Hsiang-Tsui Wang declares that acrolein is a human carcinogen. They indicate that acrolein can interact directly with DNA and induce DNA damage (Hensman et al., 1998).

In the same conditions, benzene was found 200 times higher than REL. High exposure levels of benzene, listed as carcinogen in the CMR (carcinogenic, mutagenic and reprotoxic) class make essential to reduce exposure as much as possible. Studies link benzene long-term exposure to aplastic anemia, acute leukemia, and bone marrow abnormalities. Benzene has been shown to target liver, kidney, lung, heart and the brain, causing DNA strand breaks, chromosomal damage etc (Knan, 2007). Some reactions to benzene can be immediate, such as central nervous system toxicity.

Ammonia, ethylene and toluene were found in concentrations below REL and PEL, but they are still considered toxic substances and they should be taken into consideration due to the summing effect of multiple gas exposure.

Because surgical smoke is considered to be cytotoxic, genotoxic and mutagenic [8], it is advisable to quantitatively evaluate the corresponding exposure.

However, in this study, the surgical smoke samples were analyzed without using a particle filter, so the results could be somehow altered by the presence of smoke particles inside the PA cell. Also, no considerations are made regarding the possible chemical interactions of the mentioned smoke compounds with other substances.

4.4 Carbon dioxide and water vapors detection from surgical smoke

The results of quantitative analysis of trace gas concentrations of carbon dioxide and water vapors from surgical smoke using a CO_2 laser photoacoustic system are now presented.

Surgical smoke was produced *in vitro* by irradiation of fresh animal tissue with a CO_2 laser. The tissue samples used were procured from a local company processing the meat according to the EU standards. Vaporization of tissue probes using a CO_2 laser was performed at different laser powers and different irradiation times. A gas of choice (nitrogen or synthetic air) was pumped into the cell glass through the gas inlet, while the produced smoke was transported through the gas outlet into the PA cell. Operating the PA cell at atmospheric pressure necessitates approximately 1000 cm³ of flow gas in order to fill the internal volume of the cell. Hence, our measurements correspond to a dilution of the smoke sample in 1 dm³ of buffer gas (nitrogen or synthetic air). Before reaching the PA cell, surgical smoke passes through a filter that retains particles with a diameter higher than 10 μm. Measurements were realized at the atmospheric pressure (1024-1034 mbar) and room temperature. The KOH scrubber was not used in this case.

The water vapors are a vehicle for other components (Al Sahaf et al., 2007). Water vapor itself is not harmful, but acts as a carrier, and easily passes through the smoke evacuator and filter. In our measurements, water vapor percentage ranges between 1 and 11% (see Table 7), with lower concentration in tissue skin. It was observed that the percentage of water from surgical smoke increases with the laser power and the exposure time and depends on the type of tissue.

Like any other combustion, surgical interventions with laser produce carbon dioxide, carbon monoxide, as well as ammonia (Eickmann et al., 2011). These substances are respiratory tract irritants and can cause effects related to tissue hypoxia. The carbon dioxide found in these measurements after CO_2 laser vaporization of fresh animal tissue was in high concentration, in the range of 1.34% ÷ 8.6% (see Table 7). It was observed an increase of concentration with the exposure time, in both nitrogen and synthetic air. High concentration of carbon dioxide in confined areas can be potentially dangerous. Carbon dioxide may act as an oxygen displacer and causes a number of reactions. These reactions may include dizziness, disorientation, and suffocation. An increase in inhaled carbon dioxide and subsequent reaction with water in the blood form carbonic acid (H_2CO_3) which dissociates into hydrogen ions (H^+) and bicarbonate (HCO_3) and the dissociation of carbonic acid increases the acidity of the blood (low pH) (Eickmann et al., 2011, Al Sahaf et al., 2007).

Table 7. Quantitative results of carbon dioxide and water in surgical smoke obtained from CO_2 laser ablation of fresh animal tissue

Tissue	Laser power- P [W]	Exposure time- t [sec]	CO₂ [%]	H₂O [%]	Atmosphere
Spleen, pig	6	12	3.2	3.3	synthetic air
Spleen, pig	6	20	8.4	11	synthetic air
Spleen, pig	6	12	2.5	2.8	synthetic air
Spleen, pig	6	20	4.0	8.2	synthetic air
Spleen, pig	6	20	5.8	6	nitrogen
Lung, pig	6	12	1.34	1.9	nitrogen
Lung, pig	10	6	1.15	8	nitrogen
Liver, pig	10	3	2.43	3.7	nitrogen
Liver, pig	15	3	5.8	6.4	nitrogen
Skin, pig	15	3	2.48	1.6	nitrogen
Liver chicken	15	3	6.86	3.15	nitrogen
Kidney, pig	15	3	6.16	3	nitrogen
Skin, pig	15	3	8.6	1	nitrogen

From these results it can be observed that the water vapors percentage depends also on the type of tissue. For example, the skin, which is a biological tissue with a low percent of water, has a lower percent of water vapors in surgical smoke

Conclusions

Over the last years, the smoke composition has been intensively studied in order to provide useful information for risk assessment. While the qualitative composition is well elaborated, the quantitative composition of surgical smoke still needs more data.

The present study has provided quantitative information about the smoke composition using CO_2 laser photoacoustic spectroscopy. This method, with high sensitivity and selectivity, proved once again its efficiency in gas absorption measurements. By irradiating fresh animal tissues in vitro with a CO_2 laser, it was observed that important gases like acetonitrile, acrolein, ammonia, benzene, ethylene, and toluene are present in all samples in each measurement. The results showed that there is a strong relation between smoke composition and laser power, exposure time and tissue type.

Gas concentrations were found in the ppm range for acetonitrile, acrolein, ammonia, benzene, and toluene, and in the ppb range for ethylene. Acrolein and benzene, known carcinogens, were found in concentrations exceeding the REL and PEL. Acetonitrile, a chemical that has adverse health effects was found in concentrations even hundreds of time higher than REL. Ammonia, ethylene and toluene were found in concentrations below REL a PEL, but they contribute to the total effect of multigas exposure, taking into consideration that 3 gases are already exceeding the permissible exposure limits.

Although the particle distribution, the post-irradiation chemical stability and a series of other known smoke compounds were not referenced in this study, however the targeted results are reliable thanks to the very sensitive instrument employed for detection.

The conclusion drawn from the results of this study is that the presence of surgical smoke in the operating theatres presents a real health hazard, so the recommendation of wearing higher quality filter masks is a compulsory measure to protect against surgical smoke. Although there are a variety of surgical masks that proved their filter performance, none of them can be 100% efficient.

It is recommended that masks be changed often and other protective methods must be used like proximity suction systems for smoke evacuation, OR ventilation, and wall suction systems. Involving other laser sources offering detection capacity for an increased number of gases, analyzing the breath samples of the OR personnel and patients, or taking into account the spreading dynamics of the smoke particles in the OR, more studies can be developed in order to completely identify the quantitative composition of surgical smoke and the induced health risk for developing diseases or metabolic disorders in the body.

A quantitative analysis of carbon dioxide and water vapors released in surgical smoke obtained after CO_2 laser vaporization of different fresh animal tissue was also investigated with a CO_2 laser photoacoustic system. Future work is intended in order to establish the correct correlation of carbon dioxide and water quantities with the vaporization conditions (laser power, interaction time). Water vapor is the main component of surgical smoke, although it depends on the nature of the tissue, while carbon dioxide concentration depends of laser power and exposure time. The results may differ depending on laser power, exposure time, tissue selected and environment. The values for carbon dioxide and water vapor concentrations were obtained from direct measurements inside the PA cell and not from open spaces, where gas concentrations from the surgical smoke is lower owing to dilution in air.

References

Al Sahaf O.S. , Vega-Carrascal I., Cunningham F.O., McGrath J.P., Bloomfield F.J. (2007). Chemical composition of smoke produced by high- frequency electrosurgery, *Irish Journal of Medical Science*, Vol. 176, pp. 229-232

Al Sahaf O.S., Vega-Carrascal I., Cunningham F.O., McGrath J.P., and Bloomfield F.J. (2007).Chemical composition of smoke produced by high-frequency electrosurgery, *Ir. J. Med. Sci.* Vol.176, No.3, pp. 229-232

Alp E., Bijl D., Bleichrodt R.P., Hansson B., and Voss A., (2006). Surgical smoke and infection control, *J. Hosp. Infec.* Vol. 62, No.1, pp. 1-5

Barrett W.L., and Garber S.M. (2003). Surgical smoke: a review of the literature, *Surg Endosc.* Vol. 17, No.6, pp. 979-987

Bratu A.M., Petrus M., Patachia M., and Dumitras D.C. (2013). Carbon dioxide and water vapors detection from surgical smoke by laser photoacoustic spectroscopy, *UPB Sci Bull, series A.* Vol. 75, No. 2, pp. 139-146

Chung Y.J., Lee S.K., Han S.H., Zhao C., Kim M.K., Park S.C., and Park J.K. (2010). Harmful gases including carcinogens produced during transurethral resection of the prostate and vaporization, *Int J Urol,* Vol. 17, No. 11, pp. 944-949

Cleverson Gaspareto J., Pontarolo R., and Martins Guimarães de Francisco T. (2012). Acetonitrile:properties, exposure, metabolism and toxicity, *Nova Science Publishers*

Dumitras D.C., Banita S., Bratu A.M., Cernat R., Dutu D.C.A., Matei C., Patachia M., Petrus M., and Popa C.(2010). Ultrasensitive CO$_2$ laser photoacoustic system, *Infr Phy & Tech*, Vol. 53, No.5, pp. 308-314

Dumitras D.C., Dutu D.C.A., Matei C., Magureanu A.M., Petrus M., and Popa C. (2007). Laser photoacoustic spectroscopy: principles, instrumentation, and characterization, *J Optoelectron Adv Mater*, No.9, pp. 3655-3701

Eickmann U., Falcy M., Fokuhl J., Ruegger M. (2011). Surgical smoke: Risks and preventive measures, *International Section of the ISSA prevention of occupational risks and health services*, Germany

Feng Z., Hu W., Hu Y., and Tang M.S. (2006). Acrolein is a major cigarette-related lung cancer agent: Preferential binding at p53 mutational hotspots and inhibition of DNA repair, *Proc Nat Acad Sci* USA. Vol. 103, No. 42, pp. 15404–15409

Garden J.M., O'Bannion M.K., Bakus A.D., and Olson C. (2002). Viral disease transmitted by laser-generated plume (aerosol), *Arch Dermatol*. Vol. 138, No. 10, pp. 1303-1307

Garden J.M., O'Bannion M.K., Shelnitz L.S., Pinski K.S., Bakus A.D., Reichmann M.E., Sundberg J.P. (1988). Papillomavirus in the vapor of carbon dioxide laser-treated verrucae, *JAMA*, Vol. 259, No. 8, pp. 1199-1202

Gatti J.E., Bryant C.J., Noone R.B., and Murphy J.B. (1992). The mutagenicity of electrocautery smoke, *Plast Reconstr Surg*. Vol. 89, No.5, pp. 781-784

Gianella M. and Sigrist M.W. (2010). Infrared spectroscopy on smoke produced by cauterization of animal tissue, *Sensors (Basel)*, Vol. 10, No. 4, pp. 2694-708

Gianella M. and Sigrist M.W. (2012). Chemical analysis of surgical smoke by infrared laser spectroscopy, *App Phys B*. Vol. 109, No.3, pp. 485-496

González-Bayón L., González-Moreno S., and Ortega-Peréz G. (2006). Safety considerations for operating room personnel during hyperthermic intraoperative intraperitoneal chemotherapy perfusion, *Eur J Surg Oncol*. Vol. 32, No. 6, pp. 619-624

Hensman C., Barry D., Willis R.G., and Cuschieri A. (1998). Chemical Composition of smoke produced by high-frequency electrosurgery in a closed gaseous environment-an in vitro study, *Surg Endosc*. Vol.12, No.8, pp. 1017-1019

http://www.cdc.gov/niosh/docs/97-100/pdfs/97-100.pdf

http://www.epa.gov/iris/subst/0364.htm [March 2003]

http://www.epa.gov/iris/toxreviews/0205tr.pdf

https://www.osha.gov/dte/grant_materials/fy09/sh-19495 09/health_hazards_workbook.pdf

Knan H.A. (2007). Benzene's toxicity: a consolidated short review of human and animal studies, *Hum Exp Toxicol*, Vol. 26, No. 9, pp. 677-685

link:ofmpub.epa.gov/eims/eimscomm.getfile?p_download_id=428659 [April 1998].

Moot A.R., Ledingham K.M., Wilson P.F., Senthilmohan S.T., Lewis D., Roake J., and Allardyce R..(2007). Composition of volatile organic compounds in diathermy plume as detected by selected ion flow tube mass spectrometry, *ANZ J Surg.* Vol. 77, No. 1-2, pp. 20-23

Ott D.E., Moss E., and Martinez K. (1998). Aerosol exposure from an ultrasonically activated (harmonic) device, *J. Am. Assoc. Gyn. Lap.* Vol. 5, No. 1, pp. 29-32

Petrus M., Matei C., Patachia M., and Dumitras D.C. (2012). Quantitative in vitro analysis of surgical smoke by laser photoacoustic spectroscopy, *J Optoelectr Adv Mat*, Vol. 14, No. 7-8, pp. 664-670

Sawchuk W.S., Weber P.J., Lowy D.R., and Dzubow L.M. (1989). Infectious papilloma-virus in the vapour of warts treated with carbon dioxide laser or electrocoagulation: detection and protection. *J Am Acad. Dermatol.* Vol. 21, No.1, pp. 41-49

Travella M.J., Viego J., Luiszer F., Drexler J., Blackburn P., Hovland P., and Repine J.E. (2001). Respirable particles in the excimer laser plume, *J Cataract Refract Surg.* Vol. 27, No. 4, pp. 604-607

Ulmer B. (2008). The hazard of surgical smoke, *AORN Journal*, Vol. 87, No.4, pp. 721-738

Weston R., Stephenson R.N., Kutarski P.W., and Parr N.J. (2009). Chemical Composition of Gases Surgeons Are Exposed to During Endoscopic Urological Resections, *Urology*, Vol. 74, No. 5, pp. 1152-1154

Printed by Books on Demand GmbH, Norderstedt / Germany